SpringerBriefs in Reproductive Biology

SpringerBriefs in Reproductive Biology is an exciting new series of concise publications of cutting-edge research and practical applications in Reproductive Biology. Reproductive Biology is the study of the reproductive system and sex organs. It is closely related to reproductive endocrinology and infertility. The series covers topics such as assisted reproductive technologies, fertility preservation, in vitro fertilization, reproductive hormones, and genetics, and features titles by the field's top researchers.

More information about this series at http://www.springer.com/series/11053

Denny Sakkas • Mandy G Katz-Jaffe
Carlos E Sueldo

Gamete and Embryo Selection

Genomics, Metabolomics and Morphological
Assessment

Denny Sakkas
Boston IVF Inc.
Waltham
Massachusetts
USA

Carlos E Sueldo
Univ. California San Francisco-Fresno,
Department of Obstetrics and Gynecology,
Fresno,
California
USA

Mandy G Katz-Jaffe
National Foundation for Fertility Research
Lone Tree
Colorado
USA

ISSN 2194-4253   ISSN 2194-4261 (electronic)
ISBN 978-1-4939-0988-9   ISBN 978-1-4939-0989-6 (eBook)
DOI 10.1007/978-1-4939-0989-6
Springer New York Heidelberg Dordrecht London

Library of Congress Control Number: 2014941944

Printed on acid-free paper

Springer is part of Springer Science+Business Media (www.springer.com)

Contents

List of Authors

Denny Sakkas   Boston IVF, Waltham, MA, USA

Mandy G Katz-Jaffe   National Foundation for Fertility Research, Lone Tree, CO, USA

Colorado Center for Reproductive Medicine, Lone Tree, USA

Carlos E Sueldo   Univ. California San Francisco-Fresno, Department of Obstetrics and Gynecology, Fresno, CA, USA

Morphological and Metabolic Assessment of Oocytes and Embryos

Denny Sakkas

Introduction

The non-invasive assessment of preimplantation embryos has been largely limited to the use of morphology and has become the primary tool of the embryologist for selecting which embryo(s) to replace. Since the early years of in vitro fertilization (IVF) it was noted that embryos cleaving faster and those of better morphological appearance were more likely to lead to a pregnancy. Indeed, Edwards and colleagues noted only a few years after the birth of Louise Brown "that cleavage rates on a certain day and overall embryo morphology were valuable in choosing which embryo to transfer" [1]. In 1986, one of the initial large studies ($N = 1,539$ embryos) examining the usefulness of embryo morphology was published by Cummins et al. [2] and reported that embryo quality scores were valuable in predicting success. Indeed Cummins et al. [2] calculated an embryo development rating based on the ratio between the time at which embryos were observed at a particular stage after insemination and the time at which they would be expected to reach that stage of a hypothetical "ideal" growth rate with a cell cycle length of 11.9 h. Using this scoring system, "normally" growing embryos scored 100, however the scoring system was evidently never assessed prospectively. The following year a study by Puissant et al. [3] reported the grading of embryos based on the amount of anucleate fragments expelled during early cleavage and on developmental speed. They found that embryos endowed with a high score were more often associated with pregnancy and in particular with the occurrence of multiple pregnancy. Interestingly, they already proposed that in the event of a high score: "It might be warranted to replace only two embryos when these conditions are fulfilled." Here already, in the 1980s, the simple but important concept was introduced that identifying a better embryo will allow us to transfer fewer embryos.

In addition to the classical parameters of cell number and fragmentation, numerous other characteristics have now been examined including: pronuclear

D. Sakkas (✉)
Boston IVF, 130 Second Avenue, Waltham MA 02451, USA
e-mail: dsakkas@bostonivf.com

D. Sakkas et al., *Gamete and Embryo Selection*, SpringerBriefs in Reproductive Biology,   1
DOI 10.1007/978-1-4939-0989-6_1, © Springer Science+Business Media New York 2014

morphology, early cleavage to the 2-cell stage, top quality embryos on successive days and various forms of sequential assessment of embryos (see reviews by [4–6]). One could therefore make a case that morphological assessment systems have evolved over the past decade but in effect very little has changed in the way most IVF laboratories examine embryos routinely. A close examination of the original Cummins et al. [2] paper shows that we have not really progressed in our routine assessment of morphology. One significant change however has been the ability to culture and assess blastocyst stage embryos routinely and this has helped dramatically to improve the ability to select embryos on the basis of morphology [7]. The main question is however: "How far can morphological assessment of the cleavage stage embryo go in the identification of viable embryos?"

In this review the history and progress of both morphological and metabolic assessment will be examined. The review will conclude with an evaluation of where these technologies will take us in the future.

The Changing Practice of IVF Will Challenge Classic Morphological Assessment

The drive to reduce the risk of multiple pregnancies as a consequence of IVF means that clinics around the world must transfer fewer embryos to each patient than in the past without compromising the chance of achieving a pregnancy. In order to accomplish this goal, IVF centers long used grading systems that contain semi-quantitative descriptors of the morphology of the early zygote, embryo or blastocyst (Fig. 1). Zygote grading systems evaluate pronuclear size and position, nucleoli number and distribution, and cytoplasmic appearance [8–10]. Several different criteria including the uniformity of blastomeres, percentage of fragmentation, rate of cleavage, and blastomere multinucleation are used to grade early stage cleaved embryos [10–13]. Later stage blastocyst grading systems evaluate expansion, zona thinning, and quality of the trophectoderm and inner cell mass [14, 15]. Some systems look at each stage separately and some have combined different stages and incorporate them into a "graduated" or "cumulative" embryo score [16–19].

The ultimate aim of any grading system has been to identify the zygote, embryo or blastocyst that is most likely to implant and become a healthy baby. The morphologic assessment is taken into account when deciding how many embryos to replace with the ultimate goal being a single embryo transfer (SET). Given the well-known morbidity associated with multiple pregnancy [20–22], many programs are shifting toward elective single embryo transfer (eSET). Although several countries have enacted legislation to allow the transfer of only one or two embryos [23, 24], the USA and many other countries currently have no laws to limit the number of embryos transferred but do have recommended limits issued by professional societies to encourage eSET. The American Society of Reproductive Medicine suggests that eSET is appropriate for women under age 35 with a good prognosis and a "top quality embryo" available. However, despite the recommendation the national rates

Fig. 1 An embryo development sequence taken from a real time morphology system. The real time morphology system allows continuous monitoring of the embryo without the need of removing it from the incubator. The time after insemination is annotated on the bottom right hand corner. This series depicts an embryo that progressed to the Expanded Blastocyst stage from **a** a two pronuclear embryo, **b** a two-cell embryo, **c** a four-cell embryo with some minor fragmentation, **d** a seven-cell embryo, **e** a compacted embryo and **f** an expanding blastocyst

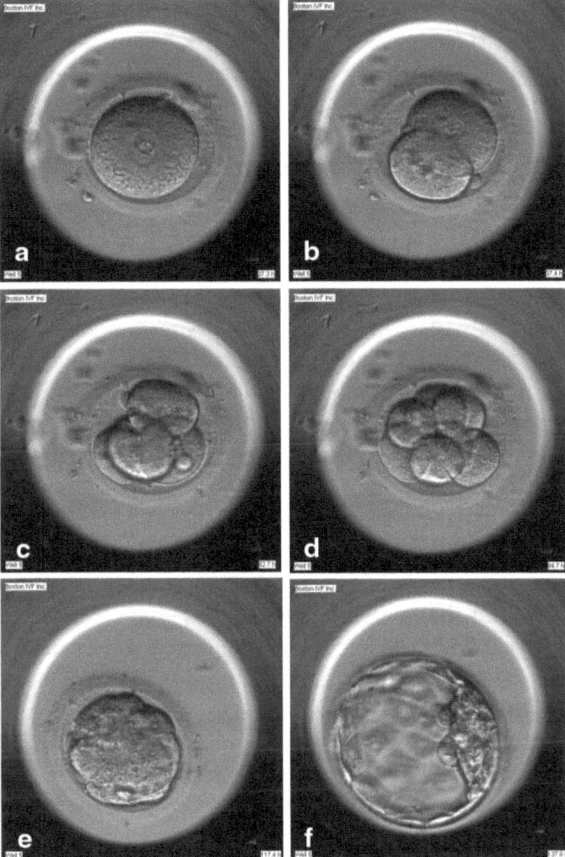

of eSET in the USA remain below 7%. In a positive sign the eSET rates are increasing in the younger age groups and are up to 11.2% in the less than 35 group [25]. The question remains for all IVF physicians and embryologists: What criteria do we use to help us pick the best embryo for transfer?

Cleavage Stage Assessment

The usefulness of morphology has been shown numerous times. Recently, a large grading study using the Society for Assisted Reproductive Technology (SART) database found that day 3 morphology was indeed useful when correlating to live birth [26]. Relationships were identified between live birth, maternal age, and morphology of transferred day 3 embryos as defined by cell number, fragmentation, and blastomere symmetry. Logistic multiple regressions and receiver operating characteristic curve analyses were applied to determine specificity and sensitivity for correctly classifying embryos as either failures or successes. Live birth rate was positively associated with increasing cell number up to eight cells (<6 cells: 2.9%; 6 cells: 9.6%; 7 cells: 15.5%; 8 cells: 24.3%; and >8 cells: 16.2%), but was negatively associated with maternal age, increasing fragmentation, and asymmetry scores. An area under the receiver operating curve (AUC) of 0.753 (95% confidence interval 0.740–0.766) was derived, with a sensitivity of 45.0%, a specificity of 83.2%, and 76.4% of embryos being correctly classified with a cutoff probability of 0.3.

Interestingly, when similar models were applied to some sequential scoring systems they appear to not have helped as much as expected. Models built using Day 1, 2 or 3 scores independently on the re-sampled data sets showed that Day 1 evaluations provided the poorest predictive value (median AUC = 0.683 versus 0.729 and 0.725, for Day 2 and 3). Combining information from Day 1, 2 and 3 marginally improved discrimination (median AUC = 0.737). Using the final Day 3 model fitted on the whole dataset, the median AUC was 0.732 (95% CI, 0.700–0.764), and 68.6% of embryos would be correctly classified with a cutoff probability equal to 0.3. The authors concluded that Day 2 or Day 3 evaluations alone are sufficient for morphological selection of cleavage stage embryos. The derived regression coefficients can be used prospectively in an algorithm to rank embryos for selection. It could be argued however that when static morphological systems are challenged with SET they will struggle to be as predictive. The usefulness of some sequential systems has been shown by Belgian groups which developed characteristics that constituted a "top quality" embryo [19–22] and showed improvement when transferring one embryo only.

The most impressive static morphology based selection results have been reported using the blastocyst scoring systems. A number of these have been developed but the most widely used system is that referred to as the Gardner Blastocyst Alphanumeric Scoring System.

The Blastocyst

It could be argued that the best static morphological selection tool available to us has been right under our noses all along [27–29]. For an embryo to form a blastocyst in culture it has already been challenged by the in vitro environment and is also a complete expression of the embryos ability to develop distinct tissue types and proceed through embryonic genome activation, reflecting both maternal and paternal genome expression. Selection of embryos up to the 8-cell stage is not always reflective of these challenges. In contrast, many laboratories argue that it is difficult to culture embryos to the blastocyst stage and also many patients fail to have blastocysts for transfer. This however is not the experience of all laboratories as demonstrated by Marek et al. [30] whereby the cancellation rates for transfer after retrieval for day 3 compared with day 5 transfer were 2.9 vs 6.7%, respectively. Another program also found that there was no difference in the percent of patients not having an embryo transfer on day-5 (2.8%) compared to day-3 (1.3%) [31]. Both studies, more than 10 years ago, concluded that using extended embryo culture in a nonselective manner for couples undergoing IVF was feasible.

A propensity of studies has shown that blastocyst transfer is more successful than transfer of cleavage stage embryos. The most recent Cochrane Data base analysis [32, 33] has shown that there was evidence of a significant difference in implantation rate and live birth rate per couple favoring blastocyst culture. The most recent report showed that in 1510 women the Live Birth Rate was 31% for Day 2–3 and 38.8% for Day 5–6. Although this report did not show a difference in cumulative pregnancy rates, it would be expected that more up to date data will also lead to improvements in cryopreservation of blastocysts as more vitrification data is published [34]. This data indicates that vitrified blastocysts are virtually equal to fresh blastocysts in their viability [34].

In order to select the best blastocyst for transfer, in humans, three morphological parameters have routinely been used, i.e. degree of blastocoele expansion and appearance of both the trophectoderm (TE) and the inner cell mass (ICM) (Fig. 1). Although it has been shown that blastocysts with highest scores for all three parameters achieve highest implantation rates, their independent ability to predict pregnancy outcome has recently come under scrutiny. Ahlstrom et al. [35] performed a retrospective analysis of 1117 fresh day 5 single blastocyst transfers and examined the live birth outcome related to each morphological parameter. Whereas all three morphological characteristics had a significant effect on live birth however, once adjusted for known significant confounders, it was shown that TE was the only statistically significant independent predictor of live birth outcome. They concluded that a strong TE layer is essential at this stage of embryo development, allowing successful hatching and implantation. The final barrier to performing routine blastocyst culture was the ability to cryopreserve them successfully. This has now been put to rest with the success of vitrification where success rates are being reported equivalent to fresh transfers [34]. The added benefit may also be that transferring on frozen cycles, as compared to fresh stimulated cycles, may convey further benefits to the developing fetus such as improved weight at live birth [36].

Real Time Morphology

Since the late 1980s numerous groups examined the possibility of time lapse video imaging of embryos. Indeed Cohen and colleagues published a number of studies on the prognostic value of morphologic characteristics of cryopreserved embryos [37, 38], while Payne et al. [39] used video imaging to examine polar body extrusion and pronuclear formation. Hardarson et al. [40] also used video imaging to observe the internalization of cellular fragments in a human embryo. More recently, Lemmen et al. [41] used time lapse recordings to examine kinetic markers of human embryo quality in particular when cleaving from the 1 to 2 cell stage. A number of commercial time lapse systems are now on the market or being developed for the market, including the Embryoscope, Auxogyn and Primovision. One system (The Embryoscope™) is a combined incubator and time-lapse system and has had numerous publications indicating an equivalent or elevated clinical pregnancy rate, which was attributed to a combination of stable culture conditions and the use of morphokinetic parameters for embryo selection [42]. The time lapse system produces high quality videos with the capability of annotating each individual embryo (Fig. 1). This time lapse system looks extremely promising and some algorithms have already been developed that claim to improve pregnancy rates [43]. Interestingly the algorithms rely more on de-selecting embryos that cleave abnormally than pro-actively selecting the best embryo. A second system has also been developed which aims to assist in the early prediction of which embryo will form a blastocyst [44]. These authors have published mouse data indicating progression to the blastocyst stage can be predicted with > 93 % sensitivity and specificity by measuring three dynamic, noninvasive imaging parameters by day 2 after fertilization, before embryonic genome activation. They have now also showed predictability with Human euploid embryos using similar strategies [45]. None of the time lapse systems have however undergone a rigorous clinical trial as yet to show whether they provide an overall benefit for improving single embryo transfer pregnancy rates. This data is eagerly anticipated. The real time imaging systems could however provide other benefits including the ability to monitor the embryos without removing them from the incubator. This more stable and consistent culture will limit changes in temperature and pH that the embryo experiences when being manipulated and examined outside the incubator.

Embryo Metabolism as a Means of Assessing Viability

Glucose

In 1980, Renard et al. [46], observed that Day 10 cattle blastocysts which had an elevated glucose uptake developed better, both in culture and in vivo after transfer than those blastocysts with a lower glucose uptake. Numerous studies have since validated this original observation in different species. In 1987, using non-invasive microfluorescence, Gardner and Leese [47] measured glucose uptake by individual Day-4 mouse blastocysts prior to transfer to recipient females. Those embryos that went to term had a significantly higher glucose uptake in culture than those embryos that failed to develop after transfer. Similar studies have validated this technology and shown that the glycolytic rate of mouse blastocysts could also be used to select embryos for transfer prospectively [48]. Interestingly this study only examined morphologically identical mouse blastocysts with equivalent diameters and rated them according to metabolic criteria, as either "viable" or "non viable" prior to transfer. Those selected as viable had a fetal development of 80 % while embryos that exhibited an abnormal metabolic profile (compared to in vivo developed controls), developed at a rate of only 6%. Clearly, this data provides unequivocal evidence that glucose metabolism is linked to embryo viability.

Recently, Gardner et al. [49] determined that glucose consumption on Day 4 by human embryos was twice as high in those embryos that went on to form blastocysts. They also found that blastocyst quality affected glucose uptake. Poor quality blastocysts consumed significantly less glucose than top scoring embryos. In studies on amino acid turnover by human embryos, Houghton et al., [50] determined that alanine release into the surrounding medium on Day 2 and Day 3 was highest in those embryos that did not form blastocysts. Brison et al. [51] have reported that changes in concentration of amino acids in the spent medium of human zygotes cultured for 24 h in an embryo culture medium containing a mixture of amino acids using High Performance Liquid Chromotography. They found that asparagine, glycine and leucine were all significantly associated with clinical pregnancy and live birth. Unfortunately we are still waiting for an easy to use methodology to assess these parameters. The studies performed on nutrient uptake and the subsequent viability of the human embryo have all used techniques that are still difficult to use routinely. The problem of adapting more difficult laboratory techniques to measure metabolism has led to the question of how else can the metabolic profile of an embryo be investigated?

Another approach that has been examined is one that performs a more systematic analysis of the inventory of metabolites that are present in the media an embryo is cultured in. One drawback of using this approach is that one needs to create an algorithm that relates to embryo function, whereas the other approach relies more on a candidate metabolite assessment.

Metabolomics

The complete array of small-molecule metabolites that are found within a biological system constitutes the metabolome and reflects the functional phenotype [52]. Metabolomics, is the systematic study of this dynamic inventory of metabolites, as small molecular biomarkers representing the functional phenotype in a biological system. Using various forms of spectral and analytical approaches, metabolomics attempts to determine metabolites associated with physiologic and pathologic states [53]. As has been observed with the examination of individual metabolites such as glucose, investigation of the metabolome of embryos, as detected in the culture media they grow in, using targeted spectroscopic analysis and bioinformatics has also shown differences in viability of embryos. In an initial proof of principle study Seli et al. [54] established that these differences are detectable in the culture media using both Raman and Near Infrared (NIR) spectroscopy. Briefly, a statistical formula was used to assign a relative "embryo viability score"—equating to embryo reproductive potential—and it was found that this score correlated to positive or negative implantation outcomes. Interestingly when human embryos of similar morphology

Table 1 Studies examining the clinical utility of the Near Infra Red (NIR) spectrometry system indicated that although some ability was evident in improving pregnancy results it was not consistent enough. The Hardarson study examined both Day 2 and 5 single embryo transfers, the Vergouw study examined Day 3 single embryo transfers while the final two studies combined different days of transfer

Type of NIR instrument	Study type	Outcome	Morphology	Morphology plus viametrics (NIR)
Prototype Hardarson et al. [61]	Single embryo transfer	Live birth rate	Day 2: 22/83 (26.5%)	Day 2: 27/87 (31.0%)
	–	–	Day 5: 36/80 (45.0%)	Day 5: 30/77 (39.0%)
Prototype Vergouw et al. [60]	Single embryo transfer	Live birth rate	Day 3: 68/163 (41.7%)	Day 3: 61/146 (41.8%)
Commercial Economou et al. (*Unpublished*)	Double embryo transfer	Clinical pregnancy rate	8/28 (29%)	16/28 (57%)
Commercial Sfontouris et al. (62)	Multiple embryo transfer	Clinical pregnancy implantation rate	41/86 (47.7%) 66/257 (25.7%)	21/39 (53.9%) 35/102 (34.3%)

were examined using the same NIR spectral profile their viability scores varied remarkably in relation to morphology indicating that the metabolome of embryos that look similar differ significantly [55, 56].

Although numerous preliminary studies [55–59] showed a benefit of this technique they were largely based on retrospective studies and performed in a single research laboratory as distinct from a real clinical setting. Recently, a number of clinical studies have been reported using either a prototype or commercial version of the Molecular Biometrics Inc. NIR system showing inconsistent results (Table 1). The largest of these studies were performed as Randomized Clinical Trials after SET [60, 61]. All studies compared standard Morphological techniques for embryo selection versus using the NIR system to rank embryos within a cohort that had good morphology and were being selected for either transfer or cryopreservation. In the Gothenburg study [61] both day 2 and day 5 SETs were included. Although not significant, the results indicated a possible benefit of embryo selection through addition of NIR on day 2 transfer. However it failed to show any benefit for selection of day 5 SET. Interestingly, the benefits of selecting a single good quality blastocyst on day 5 have also been found to be beneficial in many other studies.

One of the underlying problems encountered in the NIR system was that the threshold of signal distinguishing between a viable and non-viable embryo was susceptible to signal noise. As a consequence this method, that had been established and cross-validated on a larger scale, proved problematic because of the technical platform itself. This was not dissimilar to the situation faced by aneuploidy screening of embryos, whereby using FISH has proved to be inadequate [63] while it now appears that modern comprehensive screening techniques are providing more consistent results [64].

It is beyond question that markers do exist in the spent embryo culture media indicative of viability. The major benefits of a non invasive type of technology is the fact that the technology can be used on spent media and the time taken to assess the samples is very short, making it possible to perform the analysis just prior to ET. So far NIR spectroscopy, when tested in stringent clinical trials, does not appear to consistently improve the chance of selecting a single embryo for a viable pregnancy and these types of technology appear to need further development before being used as an objective marker of embryo viability.

Oxygen and Reactive Oxygen Species

Other techniques have also been reported to measure metabolic parameters in culture media; however, they have yet to be diligently tested in a clinical IVF setting. These include the self-referencing electrophysiological technique, which is a non-invasive measurement of the physiology of individual cells and monitors the movement of ions and molecules between the cell and the surrounding media [65, 66]. An alternative approach measures oxygen consumption of developing embryos using a microsensor system. Interestingly, although this technology has been shown to correlate with bovine blastocyst development, it was less successful in predicting mouse embryo development [67, 68]. A more recent study has shown some benefits by examining oxygen consumption from individual embryos close to the time of transfer and showing that the oxygen consumption pattern was associated with successful implantation [69].

Some emphasis is now being placed on the relationship between reactive oxygen species (ROS) levels in culture media and the outcome of IVF cycles. This idea was first introduced by Nasr-Esfahani and Johnson in 1990 as an explanation of abnormal development of mouse embryos in vitro. In the human, a study by Bedaiwy et al. [70] has shown that increasing levels of ROS generation in Day 3 in vitro embryo culture media may have a detrimental effect on in vitro embryo growth parameters, as well as clinical pregnancy rates in IVF and ICSI cycles.

Conclusion

Analysis of embryo morphology and the development of suitable grading systems have greatly assisted in the selection of human embryos for transfer. We are however fast approaching a revolution in the way we assess embryos prior to transfer. It is highly likely that all IVF laboratories will contain some type of real time imaging system in the future which will allow both assessment of morphology and the ability to retain embryos in a constant temperature and pH without moving them for assessment. As a significant adjunct to morphology we will be using either non-invasive and/or invasive methods more routinely to help in selecting which single embryo to transfer and cryopreserve. The non-invasive analysis of embryo physiology and function using metabolic parameters will definitely be one tool that will allow us to better quantify embryo viability. The addition of such technologies will be of immense value in helping both clinicians and embryologists to more confidently select the most viable single embryo within a cohort, helping us reach the goal of all our patients to achieve pregnancy.

References

1. Edwards R, Fishel S, Cohen J. Factors influencing the success of in vitro fertilization for alleviating human infertility. J In Vitro Fert Embryo Transf. 1984;1:3–23.
2. Cummins J, Breen T, Harrison K, Shaw J, Wilson L, Hennessey J. A formula for scoring human embryo growth rates in in vitro fertilization: its value in predicting pregnancy and in comparison with visual estimates of embryo quality. J In Vitro Fert Embryo Transf. 1986;3:284–95.
3. Puissant F, Van RM, Barlow P, Deweze J, Leroy F. Embryo scoring as a prognostic tool in IVF treatment. Hum Reprod. 1987;2(8):705–8.
4. De Neubourg D, Gerris J. Single embryo transfer—state of the art. Reprod Biomed Online. 2003;7(6):615–22.
5. Sakkas D. Evaluation of embryo quality. A comprehensive textbook of assisted reproductive technology. In: Gardner D, Weissman A, Howles C, Shoham Z, editors. Laboratory and clinical perspectives. London: Martin Dunitz Press; 2001. pp. 223–232.
6. Sakkas D, Gardner DK. Noninvasive methods to assess embryo quality. Curr Opin Obstet Gynecol. 2005;17(3):283–8.
7. Gardner DK, Surrey E, Minjarez D, Leitz A, Stevens J, Schoolcraft WB. Single blastocyst transfer: a prospective randomized trial. Fertil Steril. 2004;81(3):551–5.
8. Payne JF, Raburn DJ, Couchman GM, Price TM, Jamison MG, Walmer DK. Relationship between pre-embryo pronuclear morphology (zygote score) and standard day 2 or 3 embryo morphology with regard to assisted reproductive technique outcomes. Fertil Steril. 2005;84(4):900–9.
9. Montag M, Van der Ven H, German Pronuclear Morphology Study Group. Evaluation of pronuclear morphology as the only selection criterion for further embryo culture and transfer: results of a prospective multicentre study. Hum Reprod. 2001;16(11):2384–9.
10. Weitzman VN, Schnee-Riesz J, Benadiva C, Nulsen J, Siano L, Maier D. Predictive value of embryo grading for embryos with known outcomes. Fertil Steril. 2010;93(2):658–62.
11. Ciray HN, Karagenc L, Ulug U, Bener F, Bahceci M. Early cleavage morphology affects the quality and implantation potential of day 3 embryos. Fertil Steril. 2006;85(2):358–65.
12. Hesters L, Prisant N, Fanchin R, Mendez Lozano DH, Feyereisen E, Frydman R, et al. Impact of early cleaved zygote morphology on embryo development and in vitro fertilization-embryo transfer outcome: a prospective study. Fertil Steril. 2008;89(6):1677–84.
13. Pelinck MJ, Hoek A, Simons AH, Heineman MJ, van Echten-Arends J, Arts EG. Embryo quality and impact of specific embryo characteristics on ongoing implantation in unselected embryos derived from modified natural cycle in vitro fertilization. Fertil Steril. 2010;94(2):527–34.
14. Gardner DK, Lane M, Stevens J, Schlenker T, Schoolcraft WB. Blastocyst score affects implantation and pregnancy outcome: towards a single blastocyst transfer. Fertil Steril. 2000;73(6):1155–8.
15. Balaban B, Yakin K, Urman B. Randomized comparison of two different blastocyst grading systems. Fertil Steril. 2006;85(3):559–63.
16. Terriou P, Sapin C, Giorgetti C, Hans E, Spach JL, Roulier R. Embryo score is a better predictor of pregnancy than the number of transferred embryos or female age. Fertil Steril. 2001;75(3):525–31.
17. Fisch JD, Rodriguez H, Ross R, Overby G, Sher G. The graduated embryo score (GES) predicts blastocyst formation and pregnancy rate from cleavage-stage embryos. Hum Reprod. 2001;16(9):1970–5.
18. Sjoblom P, Menezes J, Cummins L, Mathiyalagan B, Costello MF. Prediction of embryo developmental potential and pregnancy based on early stage morphological characteristics. Fertil Steril. 2006;86(4):848–61.
19. Neuber E, Rinaudo P, Trimarchi JR, Sakkas D. Sequential assessment of individually cultured human embryos as an indicator of subsequent good quality blastocyst development. Hum Reprod. 2003;18(6):1307–12.

20. Jones HW. Multiple births: how are we doing? Fertil Steril. 2003;79(1):17–21.
21. Luke B, Brown MB, Nugent C, Gonzalez-Quintero VH, Witter FR, Newman RB. Risk factors for adverse outcomes in spontaneous versus assisted conception twin pregnancies. Fertil Steril. 2004;81(2):315–9.
22. Gleicher N, Barad D. Twin pregnancy, contrary to consensus, is a desirable outcome in infertility. Fertil Steril. 2009;91(6):2426–31.
23. Gelbaya TA, Tsoumpou I, Nardo LG. The likelihood of live birth and multiple birth after single versus double embryo transfer at the cleavage stage: a systematic review and meta-analysis. Fertil Steril. 2010;94(3):936–45.
24. Jungheim ES, Ryan GL, Levens ED, Cunningham AF, Macones GA, Carson KR, et al. Embryo transfer practices in the United States: a survey of clinics registered with the Society for Assisted Reproductive Technology. Fertil Steril. 2010;94(4):1432–6.
25. Society for Assisted Reproductive Technology (SART). 2011 data. https://www.sartcorsonline.com/rptCSR_PublicMultYear.aspx?ClinicPKID=0
26. Racowsky C, Stern JE, Gibbons WE, Behr B, Pomeroy KO, Biggers JD. National collection of embryo morphology data into society for assisted reproductive technology clinic outcomes reporting system: associations among day 3 cell number, fragmentation and blastomere asymmetry, and live birth rate. Fertil Steril. 2011;95(6):1985–9.
27. Gardner DK, Schoolcraft WB In vitro culture of human blastocysts. In: Jansen R, Mortimer D, editors. Towards reproductive certainty: Infertility and genetics beyond. Carnforth: Parthenon Press; 1999. p. 378.
28. Gardner DK, Schoolcraft WB. A randomized trial of blastocyst culture and transfer in in-vitro fertilization: reply. Hum Reprod. 1999;14(6):1663A–1663.
29. Menezo Y, Veiga A, Benkhalifa M. Improved methods for blastocyst formation and culture. Hum Reprod 1998;13(Suppl 4):256–65.
30. Marek D, Langley M, Gardner DK, Confer N, Doody KM, Doody KJ. Introduction of blastocyst culture and transfer for all patients in an in vitro fertilization program. Fertil Steril. 1999;72(6):1035–40.
31. Wilson M, Hartke K, Kiehl M, Rodgers J, Brabec C, Lyles R. Integration of blastocyst transfer for all patients. Fertil Steril. 2002;77(4):693–6.
32. Glujovsky D, Blake D, Farquhar C, Bardach A. Cleavage stage versus blastocyst stage embryo transfer in assisted reproductive technology. Cochrane Database Syst Rev. 2012;7:CD002118.
33. Blake DA, Farquhar CM, Johnson N, Proctor M. Cleavage stage versus blastocyst stage embryo transfer in assisted conception. Cochrane Database Syst Rev. 2007;17(4):CD002118.
34. Cobo A, de los SMJ, Castello D, Gamiz P, Campos P, Remohi J. Outcomes of vitrified early cleavage-stage and blastocyst-stage embryos in a cryopreservation program: evaluation of 3,150 warming cycles. Fertil Steril. 2012;98(5):1138–46.
35. Ahlstrom A, Westin C, Reismer E, Wikland M, Hardarson T. Trophectoderm morphology: an important parameter for predicting live birth after single blastocyst transfer. Hum Reprod. 2011;26(12):3289–96.
36. Shih W, Rushford DD, Bourne H, Garrett C, McBain JC, Healy DL, et al. Factors affecting low birthweight after assisted reproduction technology: difference between transfer of fresh and cryopreserved embryos suggests an adverse effect of oocyte collection. Hum Reprod. 2008;23(7):1644–53.
37. Cohen J, Inge KL, Suzman M, Wiker SR, Wright G. Videocinematography of fresh and cryopreserved embryos: a retrospective analysis of embryonic morphology and implantation. Fertil Steril. 1989;51(5):820–7.
38. Cohen J, Wiemer KE, Wright G. Prognostic value of morphologic characteristics of cryopreserved embryos: a study using videocinematography. Fertil Steril. 1988;49(5):827–34.
39. Payne D, Flaherty SP, Barry MF, Matthews CD. Preliminary observations on polar body extrusion and pronuclear formation in human oocytes using time-lapse video cinematography. Hum Reprod. 1997;12(3):532–41.
40. Hardarson T, Lofman C, Coull G, Sjogren A, Hamberger L, Edwards RG. Internalization of cellular fragments in a human embryo: time-lapse recordings. Reprod Biomed Online. 2002;5(1):36–8.

41. Lemmen JG, Agerholm I, Ziebe S. Kinetic markers of human embryo quality using time-lapse recordings of IVF/ICSI-fertilized oocytes. Reprod Biomed Online. 2008;17(3):385–91.
42. Meseguer M, Rubio I, Cruz M, Basile N, Marcos J, Requena A. Embryo incubation and selection in a time-lapse monitoring system improves pregnancy outcome compared with a standard incubator: a retrospective cohort study. Fertil Steril. 2012;98(6):1481–9.
43. Herrero J, Meseguer M. Selection of high potential embryos using time-lapse imaging: the era of morphokinetics. Fertil Steril. 2013;99(4):1030–4.
44. Wong CC, Loewke KE, Bossert NL, Behr B, De Jonge CJ, Baer TM, et al. Non-invasive imaging of human embryos before embryonic genome activation predicts development to the blastocyst stage. Nat Biotechnol. 2010;28(10):1115–21.
45. Chavez SL, Loewke KE, Han J, Moussavi F, Colls P, Munne S, et al. Dynamic blastomere behaviour reflects human embryo ploidy by the four-cell stage. Nat Commun. 2012;3:1251.
46. Renard JP, Philippon A, Menezo Y. In-vitro uptake of glucose by bovine blastocysts. J Reprod Fertil. 1980;58(1):161–4.
47. Gardner DK, Leese HJ. Assessment of embryo viability prior to transfer by the noninvasive measurement of glucose uptake. J Exp Zool. 1987;242(1):103–5.
48. Lane M, Gardner DK. Selection of viable mouse blastocysts prior to transfer using a metabolic criterion. Hum Reprod. 1996;11(9):1975–8.
49. Gardner DK, Lane M, Stevens J, Schoolcraft WB. Noninvasive assessment of human embryo nutrient consumption as a measure of developmental potential. Fertil Steril. 2001;76(6):1175–80.
50. Houghton FD, Hawkhead JA, Humpherson PG, Hogg JE, Balen AH, Rutherford AJ, et al. Non-invasive amino acid turnover predicts human embryo developmental capacity. Hum Reprod. 2002;17(4):999–1005.
51. Brison DR, Houghton FD, Falconer D, Roberts SA, Hawkhead J, Humpherson PG, et al. Identification of viable embryos in IVF by non-invasive measurement of amino acid turnover. Hum Reprod. 2004;19(10):2319–24.
52. Oliver SG, Winson MK, Kell DB, Baganz F. Systematic functional analysis of the yeast genome. Trends Biotechnol. 1998;16(9):373–8.
53. Ellis DI, Goodacre R. Metabolic fingerprinting in disease diagnosis: biomedical applications of infrared and Raman spectroscopy. Analyst. 2006;131(8):875–85.
54. Seli E, Sakkas D, Scott R, Kwok SH, Rosendahl S, Burns DH. Non-invasive metabolomic profiling of embryo culture media using Raman and near infrared spectroscopy correlates with reproductive potential of embryos in women undergoing in vitro fertilization. Fertil Steril. 2007;88(5): 1350–7.
55. Seli E, Vergouw CG, Morita H, Botros L, Roos P, Lambalk CB, et al. Noninvasive metabolomic profiling as an adjunct to morphology for noninvasive embryo assessment in women undergoing single embryo transfer. Fertil Steril. 2010;94(2):535–42.
56. Vergouw CG, Botros LL, Roos P, Lens JW, Schats R, Hompes PG, et al. Metabolomic profiling by near-infrared spectroscopy as a tool to assess embryo viability: a novel, non-invasive method for embryo selection. Hum Reprod. 2008;23(7):1499–504.
57. Scott R, Seli E, Miller K, Sakkas D, Scott K, Burns DH. Noninvasive metabolomic profiling of human embryo culture media using Raman spectroscopy predicts embryonic reproductive potential: a prospective blinded pilot study. Fertil Steril. 2008;90(1):77–83.
58. Seli E, Bruce C, Botros L, Henson M, Roos P, Judge K, et al. Receiver operating characteristic (ROC) analysis of day 5 morphology grading and metabolomic Viability Score on predicting implantation outcome. J Assist Reprod Genet. 2011;28(2):137–44.
59. Ahlstrom A, Wikland M, Rogberg L, Barnett JS, Tucker M, Hardarson T. Cross-validation and predictive value of near-infrared spectroscopy algorithms for day-5 blastocyst transfer. Reprod Biomed Online. 2011.
60. Vergouw CG, Kieslinger DC, Kostelijk EH, Botros LL, Schats R, Hompes PG, et al. Day 3 embryo selection by metabolomic profiling of culture medium with near-infrared spectroscopy as an adjunct to morphology: a randomized controlled trial. Hum Reprod. 2012;27(8):2304–11.

61. Hardarson T, Ahlstrom A, Rogberg L, Botros L, Hillensjo T, Westlander G, et al. Non-invasive metabolomic profiling of Day 2 and 5 embryo culture medium: a prospective randomized trial. Hum Reprod. 2012;27(1):89–96.
62. Sfontouris IA, Lainas GT, Sakkas D, Zorzovilis IZ, Petsas GK, Lainas TG. Non-invasive metabolomic analysis using a commercial NIR instrument for embryo selection. J Hum Reprod Sci. 2013;6(2):133–9.
63. Mastenbroek S, Twisk M, van Echten-Arends J, Sikkema-Raddatz B, Korevaar JC, Verhoeve HR, et al. In vitro fertilization with preimplantation genetic screening. N Engl J Med. 2007;357(1):9–17.
64. Wells D, Alfarawati S, Fragouli E. Use of comprehensive chromosomal screening for embryo assessment: microarrays and CGH. Mol Hum Reprod. 2008;14(12):703–10.
65. Trimarchi JR, Liu L, Porterfield DM, Smith PJ, Keefe DL. A non-invasive method for measuring preimplantation embryo physiology. Zygote. 2000;8(1):15–24.
66. Trimarchi JR, Liu L, Smith PJ, Keefe DL. Noninvasive measurement of potassium efflux as an early indicator of cell death in mouse embryos. Biol Reprod. 2000;63(3):851–7.
67. Ottosen LD, Hindkjaer J, Lindenberg S, Ingerslev HJ. Murine pre-embryo oxygen consumption and developmental competence. J Assist Reprod Genet. 2007;24(8):359–65.
68. Lopes AS, Larsen LH, Ramsing N, Lovendahl P, Raty M, Peippo J, et al. Respiration rates of individual bovine in vitro-produced embryos measured with a novel, non-invasive and highly sensitive microsensor system. Reproduction. 2005;130(5):669–79.
69. Tejera A, Herrero J, Viloria T, Romero JL, Gamiz P, Meseguer M. Time-dependent O2 consumption patterns determined optimal time ranges for selecting viable human embryos. Fertil Steril. 2012;98(4):849–57.
70. Bedaiwy MA, Mahfouz RZ, Goldberg JM, Sharma R, Falcone T, Abdel Hafez MF, et al. Relationship of reactive oxygen species levels in day 3 culture media to the outcome of in vitro fertilization/intracytoplasmic sperm injection cycles. Fertil Steril. 2010;94(6):2037–42.

Diagnostic Techniques to Improve the Assessment of Human IVF Embryos: Genomics and Proteomics

Mandy G Katz-Jaffe

Introduction

A fundamental component of assisted reproductive technologies (ART) is the selection, from a cohort of embryos, the most competent for transfer. Principally, comprehensive morphological assessment is used to determine the embryo/s with the highest implantation potential [8, 29]. Though relatively successful, morphology-based selection has limitations, with more than 70 % of in vitro fertilization (IVF) embryos failing to implant. It is likely that this failure is due to the absence of developmentally competent embryos in a cohort as well as our inability to precisely select the most competent embryo. Therefore, the field of ART would be at a significant advantage with more precise and quantitative methods of embryo viability determination. The ability to select the most developmentally competent embryo in a cohort should improve pregnancy rates while allowing for routine single embryo transfers [41]. Improvements in platform sensitivity and cost effectiveness of omics technologies, including genomics and proteomics, has enabled the investigation of new approaches other than morphology to assess human IVF embryos.

Genomics

Chromosome aneuploidies, defined as the gain or loss of an entire chromosome, contribute to the vast majority (~70%) of pregnancy losses and congenital birth defects in both natural and ART conceptions. In fact a chromosomally aneuploid embryo or fetus will never result in a normal healthy pregnancy or baby. Advanced maternal age (AMA) is the most significant risk factor for chromosome aneuploid

M. G. Katz-Jaffe (✉)
National Foundation for Fertility Research, Lone Tree, CO, USA
e-mail: MKatz-Jaffe@FertilityResearch.org

Colorado Center for Reproductive Medicine, Lone Tree, USA

D. Sakkas et al., *Gamete and Embryo Selection,* SpringerBriefs in Reproductive Biology,   15
DOI 10.1007/978-1-4939-0989-6_2, © Springer Science+Business Media New York 2014

pregnancies. Furthermore, aneuploidy rates are higher in oocytes and embryos from women in their forties as they near the end of their reproductive lifespans [26]. Only a weak association has been observed between embryo morphology and chromosome constitution, thereby supporting the hypothesis that pre-implantation genetic diagnosis (PGD) for aneuploidy screening should improve reproductive outcomes during IVF.

PGD for Aneuploidy Screening

In the beginning, PGD for aneuploidy screening involved a blastomere biopsy from a cleavage-stage embryo with fluorescent in situ hybridization (FISH) examining a selected panel of chromosomes. The chromosomes most commonly observed in pregnancy loss and aneuploid deliveries were chosen for analysis, particularly chromosomes 13, 18, 21, X and Y. Initial retrospective studies were promising however, eleven randomized control trials (RCTs) and a meta-analysis showed no beneficial effects following FISH screening of a biopsied blastomere [23]. The lack of benefit has been attributed to a combination of concerns, specifically the high incidence of mosaicism in the cleavage stage embryo which questions the value a single cell may have as the representative of the whole embryo. Other concerns include the negative impact blastomere biopsy may have on the future developmental competence of the embryo, technical errors associated with the FISH technique itself and the lack of assessment of all 23 chromosome pairs. There was one group that was able to report a benefit from blastomere biopsy with FISH for repeated implantation failure (RIF) patients <40 years old and AMA patients 41–44 years old [31]. Utilizing blastomere FISH they conducted two randomized trials and observed a significant increase in live birth rates in both the RIF study (47.9 vs. 27.9%; $P<0.05$) and the AMA study (32.3 vs. 15.5%; $P<0.01$) with PGD for aneuploidy screening [31]. Nevertheless, it was evident that PGD for aneuploidy screening required a technique that could analyze all 23 pairs of human chromosomes.

Advances in molecular biology and cytogenetic platforms have now allowed for comprehensive chromosomal screening (CCS) or full karyotyping of biopsied material from human embryos. The first studies involved metaphase comparative genomic hybridization (CGH) [10, 34, 50] and have extended on to array based CGH [14, 40], SNP array technology [35, 37], and more recently quantitative real-time PCR [43] (Fig. 1). Independent of the molecular platform developed for CCS, the technique needs to be reliable, highly accurate, cost effective, and fully validated.

When should an Embryo be Biopsied?

There are three time points during in vitro embryonic development that allow for the biopsy of genetic material for CCS. The earliest time point involves the biopsy of the oocyte's polar bodies and has been applied by several groups including ESHRE PGS

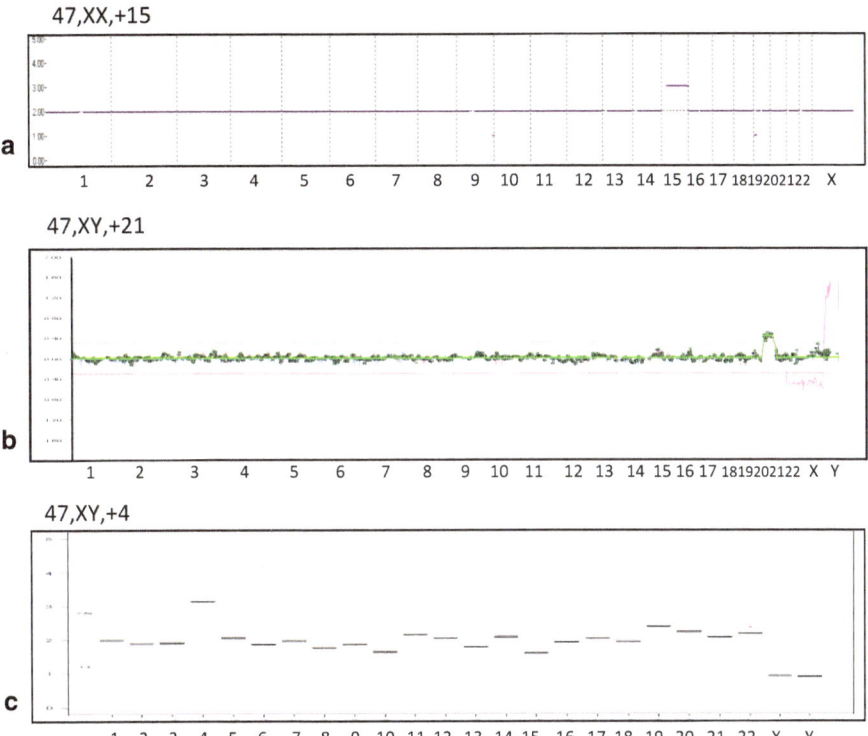

Fig. 1 a SNP array profile, **b** Array based CGH profile, and **c** Quantitative real-time PCR profile

Task Force Group [11, 12, 15, 22, 25]. Polar bodies are by products of the oocyte's meiotic divisions and are not required for either fertilization or embryonic development. Polar body biopsy is viewed as an indirect and less invasive approach of analysis of an oocyte's chromosomes (Fig. 2a). A pilot study published by the ESHRE PGS Task Force Group in AMA patients reported 76% aneuploid oocytes with a 6% discrepancy between the aneuploidy status of a polar body and the corresponding fertilized oocyte. Following embryo transfer of euploid oocytes a 30% pregnancy rate was recorded [12, 22]. Another detailed cytogenetic analysis was performed on 308 first and second polar bodies biopsied from fertilized oocytes generated by 70 infertile women of advanced maternal age (mean=40.8 years). The aneuploidy rate for this cohort of oocytes was 70%, with slightly more MII than MI errors [11]. Handyside et al. also observed over half of the aneuploidies were from the by-products of female meiosis resulting from errors in the second meiotic division [15]. In addition, they noted that most abnormal zygotes had multiple aneuploidies [15].

Interestingly, these published studies have recognized that chromosomes of all sizes participate in oocyte chromosome errors, endorsing the requirement for 23 chromosome testing. The mechanisms underlying the chromosome errors observed in oocytes include both whole chromosome non-disjunction and premature separation of sister chromatids. Further, premature separation of sister chromatids is the

Fig. 2 a Polar body biopsy, **b**
Cleavage stage biopsy, and **c**
Blastocyst biopsy

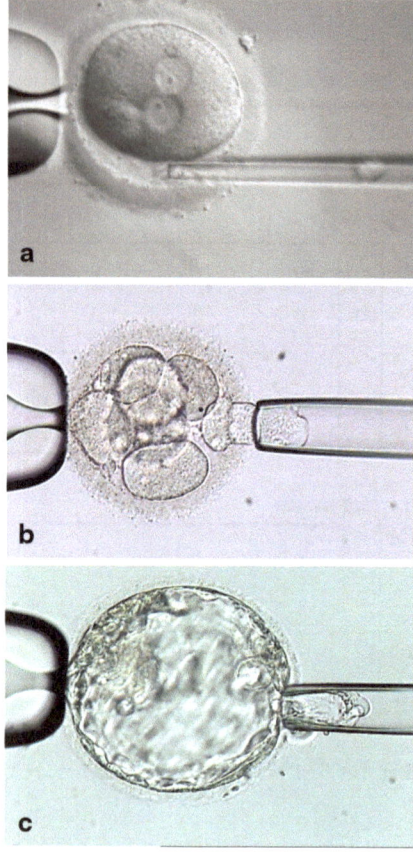

primary cause of MI errors in embryos from women with advanced maternal age
and that reciprocal loss or gain of the same chromosome typically results in euploid
embryos and even a live healthy delivery [9, 15, 38]. Nevertheless, it is important
to note that polar bodies are difficult cells to work with due to their inherent nature
to degrade which directly affects DNA quality and the coherence of chromatids.
Additionally, polar body CCS does not allow for the identification of either paternal
meiotic errors or embryonic mitotic errors. These concerns surround the clinical
utilization of polar body biopsy for CCS and whether it is a viable option for the
improvement of IVF outcome.

Embryo biopsy can involve either a cleavage stage blastomere biopsy, or a blastocyst stage trophectoderm biopsy. Cleavage stage biopsy has been to date the most
common time point for aneuploidy screening (Fig. 2b). However, recent publications question its clinical utility owing to the high incidence of moscaicism and
potential damage to the developing embryo from the biopsy procedure. A systematic review and meta-analysis of studies on embryonic chromosomal constitution
revealed 73 % of cleavage stage embryos were mosaic, with diploid-aneuploid
mosaicism the most common chromosomal constitution observed (59 %) [45].

Diploid-aneuploid mosaic embryos contain a mixture of both diploid (euploid) and aneuploid blastomeres. The accuracy and clinical viability of CCS would be significantly compromised if the biopsied blastomere did not represent the chromosome constitution of the remaining blastomeres in the embryo. A randomized and paired clinical trial investigating the impact of embryo biopsy showed an adverse effect from blastomere biopsy with 39% of cleavage stage embryos losing their ability to implant and sustain further development [39]. Only 30% of biopsied cleavage stage embryos had sustained implantation that resulted in live births, compared to 50% of unbiopsied cleavage stage embryos [39].

In contrast, it appears that limited harm is caused after biopsy at the blastocyst stage (Fig. 2c), with no measurable impact observed between biopsied and unbiopsied blastocysts in relation to IVF outcome [39]. Schoolcraft et al. observed that the probability of an individual blastocyst successfully establishing sustained implantation was 68.9%, an implantation rate 50% higher than for an individual blastocyst transferred without CCS [34]. Chromosomal mosaicism also looks to be less common at the blastocyst stage as compared with earlier embryonic stages. A recent report that reanalyzed 70 aneuploid blastocysts by isolating the inner cell mass and three segments of trophectoderm, showed a high accuracy of diagnosis with only 11 blastocysts that were mosaic (15.7%) and only two blastocysts classified as dipoid-aneuploid mosaics (2.9%) [2]. Additionally, no preferential allocation was observed of aneuploid cells between the inner cell mass and the trophectoderm [2]. Altogether, these data point towards the use of blastocyst biopsy as a practical and effective time point for chromosome screening of IVF embryos.

Evaluation of Clinical Efficacy of Comprehensive Chromosome Screening

The predictive value of embryonic reproductive potential as well as RCTs to establish efficacy are essential in order to ultimately accept the clinical validity of CCS in ART practice. A prospective, double-blinded, non-selection study using a SNP array platform for CCS was performed to measure the negative and positive predictive value of CCS in relation to embryonic reproductive potential [37]. A total of 255 IVF embryos were cultured, biopsied and selected for transfer without knowing the result of the aneuploidy screening. CCS and DNA fingerprinting were performed after transfer allowing for implantation outcome to be calculated in relation to chromosome constitution, as well as embryo to fetus identification. Results revealed that CCS was highly predictive of clinical success, with 41% of euploid embryos resulting in ongoing implantation and 96% of aneuploid predicted embryos failing to implant [37]. There are already a few published RCTs to date revealing a significant benefit using trophectoderm biopsy with CCS. One of them evaluated single embryo transfer of fresh blastocysts, with or without biopsy, in good prognosis IVF patients of young maternal age. Even in patients that did not have an increased risk for aneuploidy, the CCS group showed a significantly higher clinical pregnancy rate (69.1%) compared to the control group (41.7%) of fresh blastocyst transfer using morphology selection alone [51]. In the second RCT, IVF patients of advanced maternal age (>35 years) were randomized into either the control group of fresh blastocyst transfer based on morphology alone, or the test group of trophectoderm biopsy with CCS and blastocyst vitrification followed by a subsequent frozen embryo transfer. The ongoing clinical pregnancy rate was significantly lower at 40.9% in the control group compared to 60.8% in the CCS group [36]. Miscarriage rates were observed to be significantly lower for patients that had euploid blastocysts transferred in the CCS group compared to the control group that had traditional blastocyst morphological based selection [36]. Scott et al. [39] also showed significant improvement with CCS following fresh blastocyst transfer in a group of infertile women with a mean maternal age ~32 years. Delivery rates per cycle were significantly higher in the CCS group recorded at 84.7% compared to 67.5% for the control group without CCS [39]. These results are very encouraging and represent preliminary data of the clinical efficacy and validity of blastocyst CCS. Future completion of active RCTs involving CCS technologies is anticipated to build on this initial success, reflecting a significant improvement in the reproductive outcome for a range of infertility patients.

The future of CCS technologies lies with the rapidly developing whole genome analysis approach of next generation sequencing (NGS). NGS provides the opportunity to sequence millions of reads of DNA allowing for the simultaneous analysis of CCS and potential single-gene disorders. Recent developments in bench top sequencing platforms and sophisticated bioinformatics tools are leading the way to providing clinical CCS by NGS in the near future. A couple of publications have recently shed light on this future transition. The first study performed low coverage NGS on trophectoderm biopsies from 38 blastocysts [52]. A combination of euploid, aneuploid and structurally unbalanced blastocysts were identified by NGS and confirmed by SNP array. Only one blastocyst with different sizes of an unbalanced structural rearrangement was not confirmed by SNP array [52]. The second study published in 2014 investigated NGS for monogenic diseases on trophectoderm biopsies revealing 100% reliability with two conventional methods of single-gene mutation analysis [44]. Ongoing improvements in NGS protocols and technologies are encouraging with the potential of faster turnaround times, smaller requirements of DNA input, and less cost per run and even per embryo.

CCS goes beyond the analysis of chromosomes and allows for the development of additional markers to identify viable IVF embryos. As important as it is for all 23 pairs of chromosomes to be present for viable fetal development, not all euploid embryos, even at the blastocyst stage, will implant. Successful implantation depends on the synchronization and molecular crosstalk between a developmentally competent embryo and a receptive maternal endometrium. Any abnormality attributed to the embryo or endometrium will result in implantation failure. Utilizing other OMICS platforms, like proteomics, molecular biomarkers can be investigated to distinguish between a euploid blastocyst that has the ability to successfully implant and result in a live, healthy delivery, versus a euploid embryo that results in a negative pregnancy (Fig. 3).

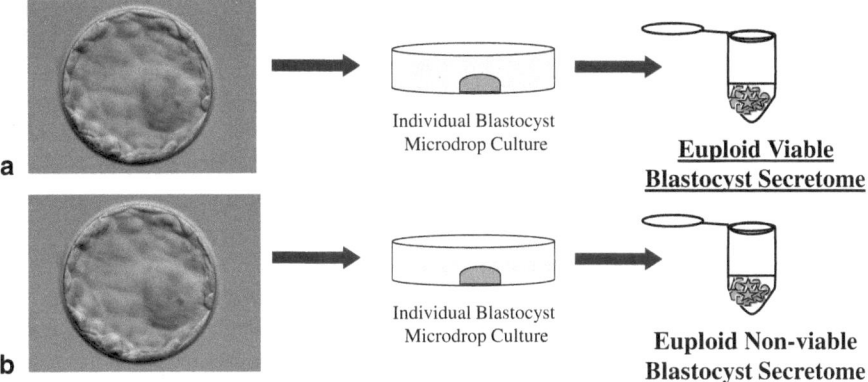

Fig. 3 Investigations to distinguish the protein secretome profiles of viable (**a**), and non-viable (**b**) blastocysts

Proteomics

The proteome represents all the proteins translated from a cell's transcriptome that are responsible for cellular function. Hence, in order to fully understand cellular function and comprehend biological processes, an investigation of a cell's proteome is vital. The proteome is complex and dynamic, constantly changing through both internal and external interactions and stimuli. The transcriptome does not always predict protein presence or abundance due to mechanisms that degrade mRNA transcripts prior to translation. Knowledge of the human oocyte and embryonic proteome is very limited even despite recent advances in proteomic technologies. The main hurdles include limited template, low protein concentration, deficient platform sensitivity, and limited protein database information.

Non-Invasive Proteomic Secretome Approach

Of particular interest to researchers trying to identify proteins involved in specific disease states is the secretome, defined as those proteins produced by cells and secreted at any given time [21]. In ART, the secretome includes those proteins that are produced by embryos and secreted into the surrounding culture medium. Analysis of the embryonic secretome would represent a non-invasive approach to embryo assessment [19]. Defining and characterizing the embryonic secretome that reflects developmental competence may improve ART outcome but will also advance our knowledge of early embryogenesis and the embryonic role during implantation [18]. To date, this has proven to be a challenging task due to the complexity and diversity of the human embryo, but holds promise with recent developments of increased sensitivity for both targeted and proteomic profiling approaches.

Early studies of the human embryonic secretome involved targeted analysis of individual proteins or molecules. The soluble factor, 1-o-alkyl-2-acetyl-sn-glyce-ro-3-phosphocholine (PAF), was one of the first molecules to be identified in the human embryonic secretome. The release of PAF influences a range of maternal physiology alterations including platelet activation and maternal immune function [28]. Leptin, a 16 kDa small pleotrophic peptide has also been observed in embryonic conditioned medium [13]. Leptin has been hypothesized to initiate and establish a molecular dialogue with leptin receptors in the maternal endometrium during the window of implantation [3]. Competent human blastocysts secrete higher leptin concentrations into the surrounding medium than arrested embryos. Another reciprocal embryo-endometrial interaction that could transform the local uterine environment, impacting both embryo development and the implantation process, is HOXA10. HOXA10 is expressed by epithelial endometrial cells and its regulation is modulated by an unknown soluble molecule secreted by human blastocysts [32].

The presence of soluble human leukocyte antigen G (sHLA-G) in embryo spent culture media has been reported in several publications to be associated with successful pregnancy outcome [20, 27]. Indeed, HLA-G has been hypothesized to also play a role during the maternal embryonic interface of implantation. However, these results have not been absolute, with pregnancies established from sHLA-G negative embryos and studies revealing undetectable levels of sHLA-G in embryo spent culture media [42, 46, 47, 49]. There are numerous factors that could influence the presence of sHLA-G in embryo spent culture media including the culture system itself, the extent of cumulus removal, single versus group embryo culture, media composition, microdrop volume and the day of media collection [47, 49]. Another explanation for the lack of reproducibility and association observed to date could be the lack of sensitivity of the current sandwich ELISA assays used for most sHLA-G analysis. It would appear that a more sensitive (picogram level) and reproducible quantitative method for analysis is required in order to determine the significance of sHLA-G in relation to embryo development and implantation outcome [47].

Human chorionic gonadotropin (hCG) is produced by trophoblast cells and is an earlier marker for the establishment of pregnancy. Investigations for the presence of intact hCG and hCG isoforms in the human embryonic secretome have revealed promising but inconsistent results. A recent study showed that intact hCG is only detected at the time point of embryo hatching and that a significant proportion of the hCG immunoreactivity is associated with hyperglycosylated hCG (hCGh), which may suggest a role with potential implantation and thus could be further investigated as a biomarker of IVF outcome [1].

Mass Spectrometry Analysis of the Embryo Secretome

The above studies have focused on only a single protein or molecule, however, it would be reasonable to assume with the complexity and multifactorial nature of embryonic development that more than one protein or molecule would be required to predict developmental competence and/or implantation potential. Mass

Spectrometry (MS) has rapidly become an important technology in proteomics research. Searching for reliable and reproducible changes in protein expression have revealed underlying molecular mechanisms of physiological processes and disease states [5]. Using Surface Enhanced Laser Desorption and Ionization MS, Katz-Jaffe et al., were the first to successfully analyze the protein secretome profile of individual human embryos [17]. The authors observed distinctive protein secretome signatures every 24 h during preimplantation development, from the time of fertilization to the blastocyst stage. Maternal proteins were observed during the first 24 h of development and unique embryonic proteins were observed in the human embryonic secretome after the activation of the embryonic genome post day 3. In addition, they reported a clear association between protein expression profiles and morphology, with degenerating embryos exhibiting significant up-regulation of several potential biomarkers that might be involved in apoptotic and growth-inhibiting pathways. Ubiquitin, a component of the ubiquitin-dependent proteosome system that is involved in a number of physiological processes including proliferation and apoptosis, was observed to have increased expression in the secretome of developing blastocysts when compared to the secretome of degenerating embryos. Secreted ubiquitin has been shown to be up-regulated in the body fluids of certain disease states and this accumulation provides evidence for an increased protein turnover [4, 33]. Ubiquitin has also been implicated in playing a crucial role during mammalian implantation by controlling the activities and turnover of key signaling molecules [48]. Two-dimensional (2D) gel electrophoresis and tandem MS have also been utilized to identify proteins in spent embryo culture media. In this study, increased levels of Apolipoprotein A1 (ApoA1) were identified in the embryonic secretome of blastocysts with higher morphological grade [53]. The presence of ApoA1 mRNA was also confirmed to be expressed in blastocysts, but not early cleavage stage embryos, suggesting that ApoA1 is a part of the embryonic transcriptome and secretome [53]. However, in relation to IVF outcome, no association was observed with ApoA1 levels in the embryonic secretome.

The incorporation of aneuploidy screening with a non-invasive method for embryo viability would be a significant advantage. Initial investigations of the blastocyst protein secretome in relation to chromosome aneuploidy have been performed using an LC-MS/MS platform. The protein profile of a euploid blastocyst secretome was markedly different from the protein profile of the aneuploid blastocyst secretome. Nine, novel, candidate biomarkers characteristically classified chromosome aneuploidy in a cohort of transferable-quality blastocysts. Lipocalin-1 was identified as the first potential biomarker for noninvasive aneuploidy screening and confirmed using an ELISA assay [24]. Lipocalin-1 has a large variety of ligands and is overproduced under conditions of stress, infection and inflammation. The increased secretion of Lipocalin-1 from an aneuploid blastocyst could represent an overall compromised state of the embryo itself that reflects the aneuploid chromosome complement. The ability to non-invasively assay for embryonic developmental competence, that included euploidy, would represent a powerful selection tool in ART.

Protein Microarray Analysis of the Embryo Secretome

Another proteomic technology that has been investigated in the characterization of the embryonic secretome is protein microarrays. In a retrospective study by Dominguez et al in 2008, protein microarrays that contained 120 targets were used to compare pooled, conditioned media from implanted versus non-implanted blastocysts following single embryo transfer [6]. Results revealed two proteins significantly decreased in the conditioned media of implanted blastocysts, CXCL13 and GM-CSF, with no proteins observed to be significantly increased. The authors hypothesized that the decrease in CXCL13 and GM-CSF indicated consumption of these proteins by the human blastocyst. Indeed, GM-CSF has been shown to promote embryo development and implantation when present in both human and murine blastocyst culture media [30].

In a subsequent study by the same group comparing protein secretome profiles between the endometrial epithelial cell (EEC) co-culture system and sequential microdrop culture media Interleukin-6 (IL-6), PLGF and BCL (CXCL13) were increased, while other proteins were decreased such as FGF-4, IL-12p40, VEGF and uPAR. IL-6 displayed the highest protein concentration in the EEC co-culture system, and upon assessment of the sequential culture media secretome using an IL-6 ELISA assay, viable blastocysts displayed an increased uptake of IL-6 compared to blastocysts that failed to result in a pregnancy, thereby suggesting a potential role for IL-6 in blastocyst development and implantation [7].

In summary, proteomic analysis is a promising technology for the development of non-invasive methods for embryo selection in ART. However, the challenge ahead still includes the reliable and reproducible identification of proteins and/or molecules associated with embryo viability and IVF success. This is a challenging task due to the complexity, heterogeneity and diversity of human embryos. Nevertheless, once these proteins and/or molecules are identified, there are sensitive, high throughput and cost effective methods available for application in an IVF clinical setting including immunodetection using ELISA or radioimmunoassays.

Conclusions

Ongoing developments in OMICS technologies are promising and are paving the way for the introduction of more quantitative, invasive and non-invasive methods for embryo selection in the field of ART. Noteworthy developments in genomics technologies, including microarrays, qPCR and NGS, have allowed comprehensive chromosome screening technologies to enter into the clinical setting and contribute to significant improvements in IVF success. The molecular assessment of the human embryonic secretome will further enhance our understanding of preimplantation embryonic development and viability. Together, the combination of a clinically proven robust quantitative non-invasive assay alongside CCS and detailed morphology assessment could represent the greatest improvement in embryo selection techniques allowing for successful routine single embryo transfers with healthy singleton deliveries.

References

1. Butler SA, Luttoo J, Freire M, Abban TK, Borrelli P, Iles RK. Human chorionic gonadotropin (hCG) in the secretome of cultured embryo: hyperglycosylated hCG and hCG-free beta subunit are potential markers for infertility management and treatment. Reprod Sci. 2013;20:1038–45.
2. Capalbo A, Wright G, Elliott T, Ubaldi FM, Rienzi L, Nagy ZP. Fish reanalysis of inner cell mass and trophectoderm samples of previously array-CGH screened blastocysts shows high accuracy of diagnosis and no major diagnostic impact of mosaicism at the blastocyst stage. Hum Reprod. 2013;28:2298–307.
3. Cervero A, Horcajadas JA, Dominguez F, Pellicer A, Simon C. Leptin system in embryo development and implantation: a protein in search of a function. Reprod Biomed Online. 2005;10(2):217–23.
4. Delbosc S, Haloui M, Louedec L, et al. Proteomic analysis permits the identification of new biomarkers of arterial wall remodeling in hypertension. Mol Med. 2008;14(7–8):383–94.
5. Dominguez DC, Lopes R, Torres ML. Proteomics: clinical applications. Clin Lab Sci Fall. 2007;20(4):245–8.
6. Dominguez F, Gadea B, Esteban FJ, Horcajadas JA, Pellicer A, Simon C. Comparative protein-profile analysis of implanted versus non-implanted human blastocysts. Hum Reprod. 2008;23(9):1993–2000.
7. Dominguez F, Gadea B, Mercader A, Esteban FJ, Pellicer A, Simon C. Embryologic outcome and secretome profile of implanted blastocysts obtained after coculture in human endometrial epithelial cells versus the sequential system. Fertil Steril. 2010;93(3):774–82. (e771).
8. Ebner T, Moser M, Sommergruber M, Tews G. Selection based on morphological assessment of oocytes and embryos at different stages of preimplantation development: a review. Hum Reprod Update. 2003;9:251–62.
9. Forman EJ, Treff NR, Stevens JM, Garnsey HM, Katz-Jaffe MG, Scott RT, et al. Embryos whose polar bodies contain isolated reciprocal chromosome aneuploidy are almost always euploid. Hum Reprod. 2013;28:502–8.
10. Fragouli E, Wells D, Whalley KM, Mills JA, Faed MJ, Delhanty JD. Increased susceptibility to maternal aneuploidy demonstrated by comparative genomic hybridization analysis of human MII oocytes and first polar bodies. Cytogenet Genome Res. 2006;114:30–8.
11. Fragouli E, Alfarawati S, Goodall NN, Sanchez-Garcia JF, Colls P, Wells D. The cytogenetics of polar bodies: insights into female meiosis and the diagnosis of aneuploidy. Mol Hum Reprod. 2011;17(5):286–95.
12. Geraedts J, et al. Polar body array CGH for prediction of the status of the corresponding oocyte. Part 1: clinical results. Hum Reprod. 2011;26:3173–80.
13. Gonzalez RR, Caballero-Campo P, Jasper M, et al. Leptin and leptin receptor are expressed in the human endometrium and endometrial leptin secretion is regulated by the human blastocyst. J Clin Endocrinol Metab. 2000;85(12):4883–8.
14. Gutierrez-Mateo C, Colls P, Sanchez-Garcia J, Escudero T, Prates R, Ketterson K, et al. Validation of microarray comparative genomic hybridization for comprehensive chromosome analysis of embryos. Fertil Steril. 2011;95:953–8.
15. Handyside AH, Montag M, Magli MC, Repping S, Harper J, Schmutzler A, et al. Multiple meiotic errors caused by predivision of chromatids in women of advanced maternal age undergoing in vitro fertilization. Eur J Hum Genet. 2012;20:742–7.
16. Hu S, Loo JA, Wong DT. Human body fluid proteome analysis. Proteomics. 2006;6:6326–53.
17. Katz-Jaffe MG, Schoolcraft WB, Gardner DK. Analysis of protein expression (secretome) by human and mouse preimplantation embryos. Fertil Steril. 2006;86(3):678–85.
18. Katz-Jaffe MG, Gardner DK. Embryology in the era of proteomics. Theriogenology. 2007;68:125–30.
19. Katz-Jaffe MG, McReynolds S, Gardner DK, Schoolcraft WB. The role of proteomics in defining the human embryonic secretome. Mol Hum Reprod. 2009;15:271–7.

20. Kotze DJ, Hansen P, Keskintepe L, Snowden E, Sher G, Kruger T. Embryo selection criteria based on morphology VERSUS the expression of a biochemical marker (sHLA-G) and a graduated embryo score: prediction of pregnancy outcome. J Assist Reprod Genet. 2010;27(6):309–16.
21. Kulasingam V, Diamandis EP. Proteomics analysis of conditioned media from three breast cancer cell lines: a mine for biomarkers and therapeutic targets. Mol Cell Proteomics. 2007;6:1997–2011.
22. Magli MC, Montag M, Koster M, Muzi L, Geraedts J, Collins J, et al. Polar body array CGH for prediction of the status of the corresponding oocyte. Part 2: technical aspects. Hum Reprod. 2011;26:3181–5.
23. Mastenbroek S, Twisk M, Van Der Veen F, Repping S. Preimplantation genetic screening: a systematic review and meta-analysis of RCTs. Hum Reprod Update. 2011;17:454–66.
24. McReynolds S, Vanderlinden L, Stevens J, Hansen K, Schoolcraft WB, Katz-Jaffe MG. Lipocalin-1: a potential marker for noninvasive aneuploidy screening. Fertil Steril. 2011;95(8):2631–3.
25. Montag M, Koster M, Strowitski T, Toth B. Polar body biopsy. Fertil Steril. 2013;100:603–7.
26. Nagaoka SI, Hassold TJ, Hunt PA. Human aneuploidy: mechanisms and new insights into an age-old problem. Nat Rev Genet. 2012;13:493–504.
27. Noci I, Fuzzi B, Rizzo R, et al. Embryonic soluble HLA-G as a marker of developmental potential in embryos. Hum Reprod. 2005;20(1):138–46.
28. O'Neill C. The role of paf in embryo physiology. Hum Reprod Update. 2005;11(3):215–28.
29. Racowsky C, Ohno-Machado L, Kim J, Biggers JD. Is there an advantage in scoring early embryos on more than one day? Hum Reprod. 2009;2:2104–13.
30. Robertson SA. GM-CSF regulation of embryo development and pregnancy. Cytokine Growth Factor Rev. 2007;18(3–4):287–98.
31. Rubio C, Bellver J, Rodrigo L, Bosch E, Mercader A, Vidal C, et al. Preimplantation genetic screening using FISH in patients with RIF and AMA: two randomized trials. Fertile Steril. 2013;99:1400–7.
32. Sakkas D, Lu C, Zulfikaroglu E, Neuber E, Taylor HS. A soluble molecule secreted by human blastocysts modulates regulation of HOXA10 expression in an epithelial endometrial cell line. Fertil Steril. 2003;80(5):1169–74.
33. Sandoval JA, Hoelz DJ, Woodruff HA, et al. Novel peptides secreted from human neuroblastoma: useful clinical tools? J Pediatr Surg. 2006;41(1):245–51.
34. Schoolcraft WB, Fragouli E, Stevens J, Munne S, Katz-Jaffe MG, Wells D. Clinical application of comprehensive chromosomal screening at the blastocyst stage. Fertil Steril. 2010;94:1700–6.
35. Schoolcraft WB, Treff NR, Stevens JM, Ferry K, Katz-Jaffe M, Scott RT Jr. Live birth outcome with trophectoderm biopsy, blastocyst vitrification, and single-nucleotide polymorphism microarray-based comprehensive chromosome screening in infertile patients. Fertil Steril. 2011;96:638–40.
36. Schoolcraft WB, Surrey E, Minjarez D, Gustofson RL, Scott RT, Katz-Jaffe MG, et al. Comprehensive chromosome screening (CCS) with vitrification results in improved clinical outcome in women >35 years: a randomized control trial. Fertil Steril. 2012;98(suppl):1.
37. Scott RT Jr, Ferry K, Su J, Tao X, Scott K, Treff NR. Comprehensive chromosome screening is highly predictive of the reproductive potential of human embryos: a prospective, blinded, nonselection study. Fertil Steril. 2012;9:870–5.
38. Scott RT, Treff NR, Stevens J, Forman EJ, Hong KH, Katz-Jaffe MG, et al. Delivery of a chromosomally normal child from an oocyte with reciprocal aneuploidy polar bodies. JARG. 2012;29:533–7.
39. Scott RT, Upham KM, Forman EJ, Zhao T, Treff NR. Cleavage-stage biopsy significantly impairs human embryonic implantation potential while blastocyst biopsy does not: a randomized and paired clinical trial. Fertil Steril. 2013;100:624–30.
40. Scriven PN, Ogilvie CM, Khalaf Y. Embryo selection in IVF: is polar body array comparative genomic hybridization accurate enough? Hum Reprod. 2012;4:951–3.

41. Seli E, Vergouw CG, Morita H, et al. Noninvasive metabolomic profiling as an adjunct to morphology for noninvasive embryo assessment in women undergoing single embryo transfer. Fertil Steril. 2010;94:535–42.
42. Tabiasco J, Perrier d'HauteriveS, Thonon F, et al. Soluble HLA-G in IVF/ICSI embryo culture supernatants does not always predict implantation success: a multicentre study. Reprod Biomed Online. 2009;18(3):374–81.
43. Treff NR, Tao X, Ferry KM, Su J, Taylor D, Scott RT Jr. Development and validation of an accurate quantitative real-time polymerase chain reaction-based assay for human blastocyst comprehensive chromosomal aneuploidy screening. Fertil Steril. 2012;97:819–24.
44. Treff NR, Fedick A, Tao X, Devkota B, Taylor D, Scott RT. Evaluation of targeted next-generation sequencing-based preimplantation gentic diagnosis of monogenic disease. Fertil Steril. 2013;99:1377–84.
45. van Echten-Arends J, Mastenbroek S, Sikkema-Raddatz B, Korevaar JC, Heineman MJ, van der Veen F et al. Chromosomal mosaicism in human preimplantation embryos: a systematic review. Hum Reprod Update. 2011;17:620–7.
46. Vercammen MJ, Verloes A, Van de Velde H, Haentjens P. Accuracy of soluble human leukocyte antigen-G for predicting pregnancy among women undergoing infertility treatment: meta-analysis. Hum Reprod Update. 2008;14(3):209–18.
47. Vercammen M, Verloes A, Haentjens P, Van de Velde H. Can soluble human leucocyte antigen-G predict successful pregnancy in assisted reproductive technology? Curr Opin Obstet Gynecol. 2009;21(3):285–90.
48. Wang HM, Zhang X, Qian D, et al. Effect of ubiquitin-proteasome pathway on mouse blastocyst implantation and expression of matrix metalloproteinases-2 and -9. Biol Reprod. 2004;70(2):481–7.
49. Warner CM, Lampton PW, Newmark JA, Cohen J. Symposium: innovative techniques in human embryo viability assessment. Soluble human leukocyte antigen-G and pregnancy success. Reprod Biomed Online. 2008;17(4):470–85.
50. Wilton L, Voullaire L, Sargeant P, Williamson R, McBain J. Preimplantation aneuploidy screening using comparative genomic hybridization or fluorescence in situ hybridization of embryos from patients with recurrent implantation failure. Fertil Steril. 2003;80:860–8.
51. Yang Z, Liu J, Collins GS, Salem SA, Liu X, Lyle SS, et al. Selection of single blastocyst for fresh transfer via standard morphology assessment alone and with array CGH for good prognosis IVF patients: results from a randomized pilot study. Mol Cytogenet. 2012;5:24–9.
52. Yin X, Tan K, Vajta G, Jiang H, Tan Y, Zhang C, et al. Massivley parallel sequencing for chromosomal abnormality testing in trophectoderm cells of human blastocysts. BOR. 2013;88:69.
53. Mains LM, Christenson L, Yang B, Sparks AE, Mathur S and Van Voorhis BJ. Identification of apolipoprotein A1 in the human embryonic secretome. Fertil Steril. 2011 96(2):422–7.

Assessment and Selection of Human Sperm for ART

Carlos E Sueldo

Introduction

Male factor is responsible for 30–40 % of all cases of human infertility. In the past, medical decisions on treating these infertile couples were based mostly on the results of the conventional semen analysis, assessing sperm concentration, motility and morphology in one or more semen samples.

During the early days of ART, cases of severe male factor were met with frustrating results, highlighted by poor fertilization and pregnancy rates. With the preliminary reports on ICSI (intracytoplasmic sperm injection) in the early 90's, clinicians and embryologists believed they had found a solution to all cases of male factor infertility, and although ICSI has become a formidable tool in ART, there are still many cases of low or absent fertilization rates despite its use, emphasizing the fact that other factors may be involved, including sperm DNA fragmentation or sperm morphologic damage that could not be detected by the standard magnification used in conventional ICSI. Lately, a number of techniques have been reported aimed at better selecting the sperm to be used in conventional ICSI, with the objectives of increasing the fertilization rate, enhancing embryo quality after successful fertilization and optimizing pregnancy rates after transfer.

Sperm DNA Fragmentation

Several studies have demonstrated the importance of the stability of the spermatic nuclei and its correlation to successful reproduction in animals and humans, and that its damage is associated to low fertilization rates, poor embryonic implantation and an increase in miscarriage rates [5, 10, 21, 27, 29, 35, 61, 66, 68, 88, 92, 105, 108].

C. E. Sueldo (✉)
Univ. California San Francisco-Fresno, Department of Obstetrics and Gynecology,
155 N. Fresno St, Fresno, CA 93701, USA
e-mail: drsueldo@hotmail.com

D. Sakkas et al., *Gamete and Embryo Selection*, SpringerBriefs in Reproductive Biology,   29
DOI 10.1007/978-1-4939-0989-6_3, © Springer Science+Business Media New York 2014

Multiple factors can induce DNA fragmentation, such as varicocele, cryptor-chidism, advanced paternal age, severe teratozoospermia, episodes of high fever, exposure to radiation or chemotherapy [15, 16, 37, 41, 57, 65, 69, 70, 99, 100, 103, 104, 106, 109], in addition to metabolic causes (lipid peroxidation and apoptosis). The presence of DNA fragmentation can also be associated with alterations in chromatin compaction or the presence of oxidative stress, either during spermatogenesis or during transit through the epididymis [3, 6–8, 20, 32, 34, 35].

The DNA damage can be present as single or double strand breaks, and both types can be analyzed and/or quantified through different methods including: **SCD** (Sperm chromatin dispersion), **SCSA** (Sperm Chromatin Structure Assay) and **TUNEL** (deoxynucleotidyl transferase-mediated dUTP nick end labeling) [31, 40, 43, 48] which will be described in more detail later in this chapter, as its presence can have a negative impact in IVF-ICSI results [42, 47].

Apoptosis (programmed cell death) is an important event in the regulation of spermatogenesis in mammals, including humans [18, 19]; in some cases, this physiologic process can be altered by various factors promoting DNA fragmentation, and in the case of sperm production these abnormal sperm would be found in the ejaculate [22, 71, 72, 81, 90, 91, 98, 106]. Apoptosis and DNA damage can be associated in some cases with alteration in the levels of radical oxygen species (ROS) [2, 4, 60, 95] or with a diminishing antioxidant capability of the seminal plasma, resulting in poor sperm quality [62, 78, 96]. ROS are highly reactive agents that belong to a class of free radicals and can be described as "any atom or molecule that has one or more unpaired electrons". They can function as mediators in the induction and development of sperm hyperactivation, capacitation and acrosomic reaction [36, 51, 56]. On the other hand, an excess of ROS results in lipid peroxidation and damage to the sperm membrane resulting in loss of motility, damage to the acrosomal membranes and DNA oxidation, leading to an inability of the sperm to fertilize the oocyte or giving origin to a non-viable pregnancy [12, 13, 44].

Sakkas and Alvarez [88] described the mechanisms by which sperm DNA can suffer structural alterations, even damaging the mitochondrial DNA. They described 6 mechanisms that can take place during sperm production or transport. Those mechanisms are: (1) apoptosis during spermatogenesis (2) DNA strand rupture during chromatin remodeling in the process of spermiogenesis (3) DNA fragmentation post-testicular induced mainly by ROS during the passage through the seminiferous tubules and/or epididymis (4) DNA fragmentation induced by endogenous caspases (5) DNA fragmentation induced by radio or chemotherapy (6) DNA damage caused by environmental toxicants. It is considered in general, that if the DNA damage is post-testicular, the levels of fragmentation would be more significant than those of testicular origin [49, 50, 77].

1. **Apoptosis during spermatogenesis**: during the process of spermatogenesis, the human testicle shows a 50–60% apoptosis rate in the different germinal lines that flow into Meiosis I. These cells are normally phagocytized by Sertoli cells, after detection by apoptotic mechanisms including the Fas receptors [89]. This mechanism is not always efficient and there is a chance that some of these "abnormal cells" enter the following steps of spermatic remodeling and therefore may be present in the ejaculate [88].

2. **DNA strand rupture during chromatin remodeling**: The process of chromatin remodeling can generate rupture of the DNA strands. It has been reported that the presence of nicks in the sperm DNA may represent an incomplete maturation [64], and if they are not repaired afterwards, would make the DNA strands more susceptible to a post-testicular damage.

3. **Post-testicular DNA fragmentation**: some men with idiopathic infertility have high levels of ROS in the semen, as well as low levels of antioxidants in comparison to fertile men. They may have infections with leukocytospermia also associated with high levels of ROS, since the white cells are the principal source of ROS in the ejaculate; but even in cases with lower concentration of white cells than the threshold dictated by the World Health Organization [1] they may cause oxidative stress. The DNA damage caused by ROS may be demonstrated by the presence of 8-OH-dG (8-hydroxy-2-deoxyguanosine) [59].

4. **DNA fragmentation caused by caspases**: It has been reported that exposure to high temperatures can induce DNA fragmentation [20]. The damage can take place in the epididymis and be caused by ROS or by the activation of spermatic caspases.

5. **DNA damage caused by radio or chemotherapy**: the exposure to radio or chemotherapy for cancer treatments can be associated with DNA fragmentation. Sometimes the damage may be associated to the disease itself, as some patients with Hodgkin's lymphoma or testicular cancer show signs of DNA fragmentation before undergoing cancer treatment [73].

6. **DNA damage caused by environmental toxicants**: Several authors have shown that air pollution, pesticides or other toxicants can cause sperm DNA damage, impacting on the patient's reproductive capacity. Performing SCSA, Evenson, (2005), showed a dose dependent damage on DNA caused by environmental toxicants.

How to Test for Human Sperm DNA Fragmentation?

In the literature there are multiple ways to detect the presence of sperm DNA fragmentation (TUNEL, SCSA, COMET Assay, SCD), or oxidative stress (8-OH 2-dG, BODIPY C11) [63].

TUNEL (Terminal Deoxynucleotidyl Transferase dUTP Nick End Labeling) This technique allows the evaluation of single or double stranded DNA damage. It measures "in situ" the presence of 3'OH free groups, which is a product of the ruptured DNA strands. It is one of the most clinically used techniques to evaluate sperm DNA damage. It is based on the fact that the enzyme "deoxinucleotidyl transferase" (TdT) allows for labeling of the DNA nicks through deoxiuridine triphosphate (dUTP). The way by which the spermatozoa are processed and the results are read, by fluorescence microscopy or flow cytometry [15, 39, 103], may create differences in results among different laboratories. The normal level is considered to be under 20% of damaged spermatozoa [93, 97] (Fig. 1). At our Center, Uriondo et al. [103] reported that in 72 patients with male infertility, there was a positive correlation between TUNEL and

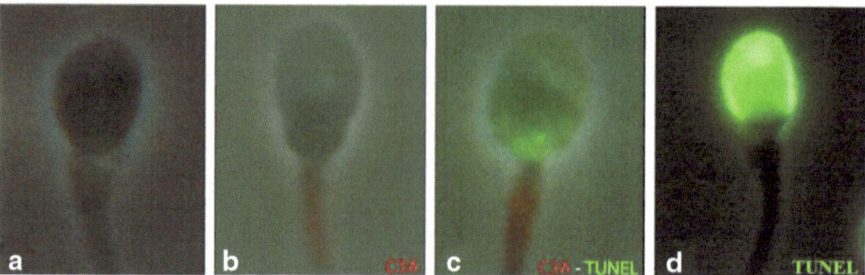

Fig. 1 TUNEL and Active Caspase 3 testing in human sperm **a** Normal, **b** AC3+/TUNEL-, **c** AC3+/TUNEL+ and **d** AC3-/TUNEL+

phosphatidylserine determinations, the values were higher in patients over 45 y/o and in those with Kruger Index under 5%, in comparison to patients without those features [11].

SCSA (Sperm Chromatin Structure Assay) In this technique, the presence of DNA fragmentation is indirectly evaluated by looking at the susceptibility of the sperm DNA to acid denaturation in situ, followed by staining with acridine orange [40]. By using flow cytometry, around 5,000–10,000 sperm can be assessed in a few seconds. Through a specific SCSA-software, the ratio between normal and abnormal sperm is determined. The percentage of red sperm (DNA fragmentation index or DFI) represent those sperm with denatured DNA, which in normal semen should be under 27–30% [26, 28].

Comet Assay This technique is based on the use of electrophoresis, and the principle that the fragmented DNA has a greater migrating velocity toward the anode of the electrophoretic field compared to the intact DNA. This assay can be done at a neutral or alkaline pH, yet this may sometimes lead to an overestimation of ruptured DNA in the sperm [94]. The normal levels reported in the literature vary among authors, [101] as there is no standardized protocol; in addition, there is a need for computerized programs with high costs involved.

SCD (Sperm Chromatin Dispersion or Halo test) [67] This technique consists in generating a differential decondensation of the chromatin between those sperm with fragmented DNA and intact DNA. This effect is achieved by using an acid treatment followed by de-proteinization, so that those sperm with fragmented DNA would not release DNA loops and do not generate a chromatin dispersion halo. Those sperm without DNA fragmentation show a big halo that corresponds to the DNA loops. It has been reported that using this technique, the chances of achieving a pregnancy is very low if 30% or more fragmentation is shown [43].

8-OH 2-dG Several studies have shown a strong association between DNA fragmentation (using TUNEL or SCSA) and oxidative stress measured by 8-OH 2-dG [5, 35]. The degree and range of oxidative stress in the sperm will depend on the nature of the agents causing the stress [46]. The main sources of ROS in semen are white cells and abnormal sperm; the leukocytes generate toxic agents that are oxygen derived. Measuring 8-OH 2-dG can be done by HPLC, Flow cytometry, ELISA and Western blot. Aitken et al. showed a high correlation between DNA fragmentation and the production of 8-OH 2–dG, establishing normal values of 40% in the whole sample and 25% for a post-gradient sample.

BODIPY C11 The oxidative stress may manifest as a lipid oxidation of the sperm plasma membrane, in what is known as lipid peroxidation. A fluorescence assay has been developed to detect the formation of lipid peroxides in bulls, pigs and humans [9]. This assay is based on determining the change of BODIPY C11 from red to green by Flow cytometry or fluorescence microscopy. Our group (Alvarez Sedó 2012) has successfully used this technique to study the effect of male aging on DNA fragmentation, and its relation to oxidative stress [15].

Sperm Morphology Assessment Beyond Kruger

A number of years ago, different investigators began to study the presence of so-called vacuoles in the sperm head. One of the first was Bartoov et al. using a system of high magnification, based on the use of inverted microscopy equipped with a Nomarski optic, a 100X objective and a system of digital amplification, allowing the observation of samples in a monitor at a final magnification of approximately 6000X or higher. This author reported that the absence of vacuoles at high magnification correlated favorably with good fertilization and clinical pregnancy rates when those sperm were selected for ICSI [23, 24, 25].

But what is the significance of finding vacuoles in the sperm head?, Franco et al. found that sperm with large vacuoles (beyond 50% of the nuclear surface) showed a level of DNA fragmentation significantly greater than those sperm with a normal nuclear morphology [45, 75, 76]. On the other hand, not all cases of increased DNA fragmentation have nuclear vacuoles, ([15], Personal Communication) therefore they are not always directly related, but they do coexist frequently.

During the process of spermiogenesis there are changes of great significance, among them the migration of mitochondria toward the base of the sperm head, the location of the Golgi apparatus in the anterior part of the nucleus and a reduction in the cytoplasmic volume. There are also changes in the synthesis and structure of different proteins, like the transition from histones to protamines and as a consequence, there is formation of various proteins that stopped being functional and ought to be eliminated. The nuclear pocket in the human sperm would be responsible for the protein degradation, the proteins to be eliminated would be labeled by the system of poly-ubiquitin, to be incorporated into the vacuoles and from there moved to the nuclear pocket where the proteasomes will cause the final protein degradation. A failure in the degradation or a structural change in some of the proteins could lead to the formation of large vacuoles and the saturation or insufficiency of the proteasomes, giving rise to the formation of large size vacuoles and the consequences previously mentioned [52, 53].

Given the potential negative impact on the reproductive results, it would be important to grade the semen samples considering these structures. The development of MSOME (motile sperm organelle morphology examination) as a diagnostic tool, consists of evaluating generally 100 sperm at high magnification, and placing them

Table 1 MSOME parameters. (Bartoov et al. 2002)

1° Selection	Oval shaped head with larger or smaller size related to normal
	Absence of vacuoles or one smaller than 4% of the nuclear area
2° Selection	No oval shaped head
	Absence of vacuoles or one smaller than 4% of the nuclear area
3° Selection	Regional disorders: extrusions or invaginations
4° Selection	Oval shaped head with larger or smaller size related to normal, but with the presence of nuclear vacuoles
5° Selection	No oval shaped head with the presence of nuclear vacuoles

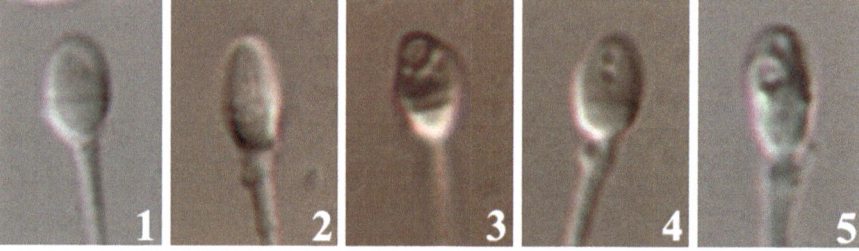

Fig. 2 Pictures of human sperm under electromicroscopy at high magnification, showing the patterns described by Bartoov et al. *1* 1st selection: oval sperm head without evidence of vacuoles, *2* 2nd selection: sperm head not oval without vacuoles, *3* 3rd selection: regional disorders (extrusions), *4* 4th selection: oval sperm head with vacuoles, *5* 5th selection: sperm head not oval with vacuoles

Table 2 A modified classification of MSOME (Motile sperm organelles morphology examination) parameters, as used in our Center

1° Selection	Sperm head without vacuoles
2° Selection	One vacuole smaller than 20% of the nuclear surface
3° Selection	Two vacuoles smaller than 20% of the nuclear surface
4° Selection	One vacuole larger than 30–50% of the nuclear surface
5° Selection	Multiple vacuoles smaller than 20% of the nuclear surface
6° Selection	Multiple vacuoles larger than 20% of the nuclear surface

in the various categories based on the findings encountered. One of the most common classifications is the one reported by Bartoov et al. Table 1, [23], Fig. 2. Other authors, including our group, introduced variants to this classification, considering mainly the size and quantity of vacuoles present in the spermatic nucleus [30]. Our group proposed a simplified classification based on the number of vacuoles and the surface area they occupy in the sperm nucleus (Table 2) Fig. 3, as we believe this modification makes the sperm assessment easier and faster. Using this classification we observed that in normozoospermic patients, the level of DNA fragmentation was significantly greater in those sperm with vacuoles that occupy more than 20% of the nucleus. We concluded that the presence of large size vacuoles alter the DNA integrity and that we did not identify differences in the externalization of phosphatidylserine, implying that said fragmentation is not produced by the apoptotic pathway.

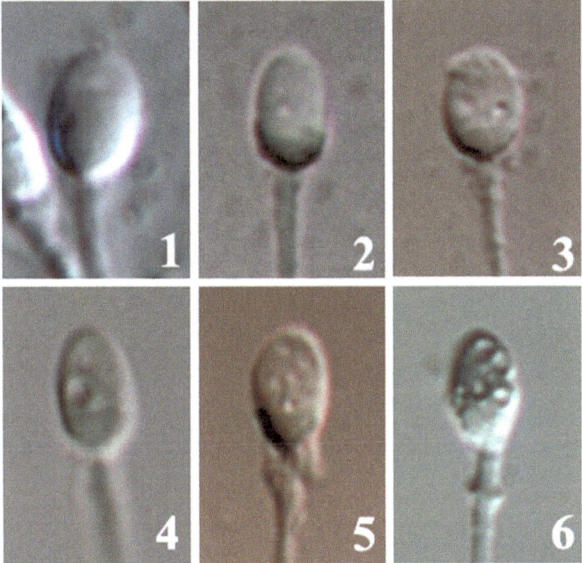

Fig. 3 Shown in this picture: *1* sperm without vacuoles, *2* small vacuole <20% of the sperm head surface, *3* two vacuoles with <20% of the sperm head surface, *4* 1 vacuole with a size >20%, *5* multiple vacuoles with <20% of the sperm head surface, *6* multiple vacuoles with >20% of the head surface

Given all the findings listed above, we propose MSOME as the preferred way to study sperm morphology in ART candidates, as we see advantages over the Kruger criteria [74]; MSOME evaluates the sperm motile fraction, which is the population that is commonly used in ART, and has more strict criteria of sperm selection since it identifies vacuoles and chromatin alterations. Furthermore, the results obtained on a given sample remain consistent over time [75, 76].

Analysis of New Techniques of Sperm Selection

Selection of Non-Apoptotic Sperm, the Use of Annexin V Columns Programmed cell death (apoptosis) is associated to controlled DNA fragmentation, which is considered the final event in this process. There are a number of apoptotic markers in ejaculated human sperm, among them: the externalization of phosphatidylserine, active caspase-3, and DNA fragmentation; this last marker is the most relevant clinically and can be assessed by TUNEL or SCSA. The ANNEXIN V column has been proposed as a relevant technique with the intent to select non-apoptotic spermatozoa, to be used in samples with high levels of apoptosis, achieving lower concentrations of DNA fragmentation, active caspase-3 and phosphatidylserine. The technique is based on the high affinity that exists between Annexin V and phosphatidylserine, a

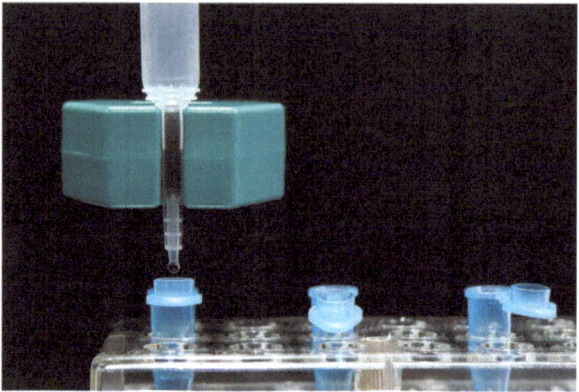

Fig. 4 Human sperm going through an annexin V column exposed to a magnetic field, at the end of the column a drop is observed falling into the collecting tube containing culture media

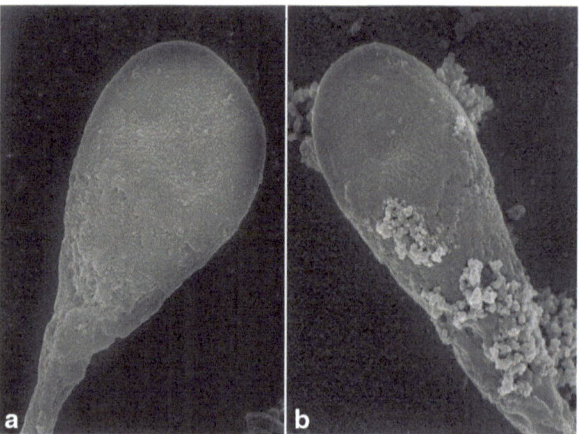

Fig. 5 Human sperm observed under electromicroscopy: without evidence of membrane binding of beads with annexin V, after going through a column exposed to a magnetic field **a** negative fraction. **b** positive fraction, showing a preapoptotic human sperm with numerous beads of annexin V in its surface

membrane phospholipid that during the early stages of apoptosis becomes externalized, migrating from the internal to the external layer.

To examine sperm samples for this apoptotic marker protein, we associate the Annexin V antibody to metal spheres (50 nm) that are co-incubated with motile sperm obtained after a separation gradient of the sample. After this period of incubation, the suspension (spheres of annexin V + sperm) is placed in a sterile column situated within a magnetic field (0, 5 T), so that those damaged sperm (apoptotic) get trapped in the matrix of the column, while those non-apoptotic sperm go through and are collected to be used in ART (Figs. 4 and 5). When to indicate Annexin V

Columns in clinical practice? At our Center we evaluate DNA fragmentation and apoptosis, by using TUNEL and Active Caspase 3 respectively [104] (Fig. 1). When the level is over 20% (TUNEL) or higher than 11% for active Caspase 3, we indicate the use of Annexin V Columns, obtaining a final sample that has a lower level of DNA fragmentation compared to the initial sample [14, 82, 86]. Rawe et al. [83] at our Center, reported the birth of a healthy newborn after reducing DNA fragmentation through Annexin V Columns, while other authors have also published similar favorable reports [87]. Dirican et al. [38], performed a controlled study (122 study patients vs 74 control patients) and showed a higher pregnancy rate after MACS (magnetic activated cell sorting). Other authors showed a beneficial effect in cases of male infertility with high DNA fragmentation, but the number of patients studied were small (Young O., Romany et al. ESHRE 2010). In summary, the magnetic columns seem to diminish the incidence of DNA fragmentation in men that initially showed elevated sperm levels, and this technique appears to be of benefit when used prior to ART; yet it is fair to mention, that there is a need for prospective randomized studies to confirm with certainty, that Annexin V Columns should be the standard of care prior to ICSI, when the semen sample shows high levels of DNA fragmentation.

Sperm Selection by Using Hyaluronic Acid (HA, PICSI) This technique is based on the concept that mature human sperm have surface receptors for hyaluronic acid. This major component of the extracellular matrix that surrounds the oocyte, makes those sperm that express HA surface receptors capable of binding to the matrix in vitro. The concept, developed after Huszar et al. [54, 55], reported that HA binding by human sperm indicated cellular maturity. This was performed by using a glass-bottomed dish covered with hyaluronate, demonstrating that non-binding sperm exhibited nuclear and cytoplasmic properties of diminished maturity; if the sample tested showed 80% or more sperm bound to the matrix, it was considered normal from a physiologic viewpoint and a manifestation of its nuclear-cytoplasmic maturity. That is why HA was proposed as a way of selecting mature sperm prior to ART procedures, although the many reports in the literature on the use of a PICSI-dish loaded with Hyaluronan and subsequently only using the bound sperm prior to ICSI revealed a variety of effects, from no benefit to major benefit, as well as only partial improvements in specific aspects, like better fertilization rate [79, 80].

A recent report from Italy by Tarozzi et al. [102] revealed that the Hyaluronan-binding assay (HBA) did not correlate with fertilization, embryo quality, implantation or pregnancy rates, but on the other hand the bound sperm showed good DNA integrity (TUNEL) and morphology. A recent publication by Ye et al. [107] revealed that HBA had a poor predictive value for sperm fertilizing ability in vitro (without ICSI), but a randomized prospective study by Colonna Worrilow et al. [33] in over 200 patients, showed similar degrees of fertilization rates but much better embryo quality in the PICSI group, which resulted in significantly higher clinical pregnancy rates. In this randomized study, the exposure of sperm to HA prior to ICSI, significantly improved embryogenesis and clinical outcome.

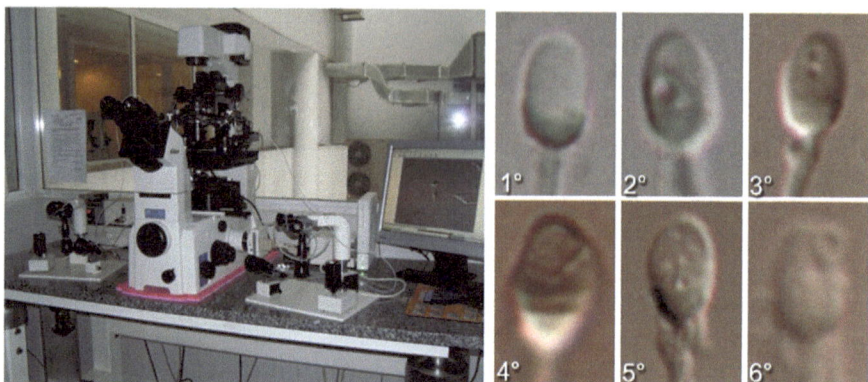

Fig. 6 (*Left*) equipment set-up at our center, for the observation of sperm at high magnification. (*Right*) showing pictures of human sperm at high magnification

Technique of Sperm Morphologically Selected Through High Magnification (IMSI) Despite the availability and use of ICSI not all cases of male infertility do well in ART. Recently, the increased use of a new technique of sperm selection prior to conventional ICSI was developed based on the previously reported data involving MSOME. It is based on the use of high magnification, obtained by using an inverted microscope with Nomarski optic, a 100X objective, and a digital system of amplification (Fig. 6), giving a final magnification higher than 6,000X. This allows for the detection of nuclear vacuoles, the presence of which have been associated to suboptimal sperm quality, since these nuclear vacuoles have been correlated with alterations in the function of the mitochondria and higher DNA fragmentation, suggesting premature chromatin decondensation.

Bartoov et al. [23] published a number of reports using this technique prior to ICSI (IMSI) and showed very good clinical results using a classification that he developed (Table 1); sperm free of vacuoles and normal shape nucleus showed better outcome than ICSI performed without prior IMSI, both in terms of better clinical pregnancy rates and lower miscarriage rates. A recent publication by Knez et al. [58] also showed beneficial effects by using IMSI prior to ICSI, demonstrating higher pregnancy rates by using IMSI vs ICSI alone in men with teratozoospermia. His findings, after injecting oocytes with sperm free of vacuoles, showed a higher number of morphologically normal zygotes and blastocysts formation rate, than when ICSI was performed with sperm having head vacuoles. These findings were confirmed in a randomized clinical trial, comparing conventional ICSI (219 patients) vs IMSI (227 patients) by Antinori et al. [17], who showed that when sperm selection in IMSI patients was performed at 6,600X, a higher rate of implantation and clinical pregnancy rates was achieved as well as lower miscarriage rate in the IMSI group, among couples with severe oligoasthenoteratozoospermia.

The technique consists of preparing a plate with a glass-bottomed dish, where the selected sperm at high magnification (using the MSOME criteria) are placed in a separate droplet, at a rate of 2–3 selected sperm per mature oocyte to be injected.

Once the sperm selection is accomplished, ICSI is subsequently carried out in a conventional manner. The sperm selection time prior to ICSI appears to be important, as the time spent by sperm in PVP may have a relation to the level of DNA fragmentation [85]. Therefore a shorter time for IMSI, after going over the learning curve and mastering the technique (helped perhaps also by using a simpler MSOME classification for sperm selection), may be ultimately improving clinical results, although this final statement has to be clinically proven by future studies.

Conclusions

It is evident that current ART procedures in cases of male infertility, cannot be carried out by simply performing a conventional semen analysis and ICSI, when one or more of the standard WHO parameters appear to be compromised. The presence of high levels of sperm DNA fragmentation or morphological nuclear abnormalities, only evident at high degrees of magnification, may be responsible for poor fertilization rates, poor embryo quality, low pregnancy rates and/or elevated miscarriage rates.

Testing for sperm DNA fragmentation is indicated in men with advanced paternal age [84], varicocele, cryptorchidism, teratozoospermia, history of poor embryo quality or unexplained repeated ART failures. As described in this chapter, there are multiple techniques available to test for DNA fragmentation, our group favors the use of TUNEL and active caspase 3, even though other methods are available. The presence of nuclear abnormalities, has shown a direct relationship with fertilization rates as well as embryo quality; these nuclear abnormalities detected and classified by performing a MSOME test at high magnification, correlated in a number of studies with the level of DNA fragmentation, although they do not always appear in tandem, as men with high sperm DNA fragmentation may not present vacuoles in the sperm nucleus.

Selecting the "best" sperm in order to perform a conventional ICSI under low magnification, as is commonly done, results in satisfactory fertilization and pregnancy rates in most cases of male infertility. As described in this chapter, a number of novel techniques have been proposed to select the "best" sperm in specific cases of male factor infertility, in order to minimize the number of sperm with apoptotic marker protein expression (MACS), to increase the odds of using for ICSI a mature sperm (PICSI) or to select sperm without nuclear abnormalities (IMSI). Large amount of clinical data has accumulated over the last few years with these techniques, mostly showing favorable results but not always, as many of the studies were poorly designed, were non-randomized, had low numbers of patients included in the studies or with findings that were not subsequently duplicated by other authors.

In our opinion, these techniques of sperm selection to be used prior to ICSI, deserve a chance in ART, as any procedure incorporated that increases the odds of using a "healthy" sperm for sperm microinjection, can only be helpful in optimizing ART outcome; as of today, some of the data accumulated is too erratic and at times poorly gathered; further confirmation by prospective randomized large scale clinical trials are needed, to more clearly define the place of these novel techniques in our armamentarium, in order to optimize ART results in this difficult group of patients with male infertility.

Acknowledgement Cristian Alvarez Sedó, Ph.D candidate, Director of Research and Mariano Lavolpe M.Sc, Senior Embryologist, at CEGYR in Buenos Aires Argentina, for their significant contributions to this chapter.

References

1. Agarwal A, Said TM. Oxidative stress, DNA damage and apoptosis in male infertility: a clinical approach. BJU Int. 2005;95(4):503–7.
2. Aitken RJ. A free radical theory of male infertility. Reprod Fertil Dev. 1994;6:19–23.
3. Aitken RJ, De Iuliis GN. Origins and consequences of DNA damage in male germ cells. Reprod Biomed Online. 2007;14:727–33.
4. Aitken RJ, Fisher H. Reactive oxygen species generation and human spermatozoa: the balance of benefit and risk. Bioassays. 1994;16:259–67.
5. Aitken RJ, Gordon E, Harkiss D, Twigg JP, Milne P, Jennings Z, Irvine DS. Relative impact of oxidative stress on the functional competence and genomic integrity of human spermatozoa. Biol Reprod. 1998;59:1037–46.
6. Aitken RJ, Koopman P, Lewis SE. Seeds of concern. Nature. 2004;432:48–52.
7. Aitken RJ, Bennetts LE, Sawyer D, Wiklendt AM, King BV. Impact of radio frequency electromagnetic radiation on DNA integrity in the male germline. Int J Androl. 2005;28:171–9.
8. Aitken RJ, Wingate JK, De Iuliis GN, Koppers AJ, McLaughlin EA. Cis-unsaturated fatty acids stimulate reactive oxygen species generation and lipid peroxidation in human spermatozoa. J Clin Endocrinol Metab. 2006;91:4154–63.
9. Aitken RJ, Wingate JK, De Iuliis GN, McLaughlin EA. Analysis of lipid peroxidation in human spermatozoa using BODIPY C11. Mol Hum Reprod. 2007;13(4):203–11.
10. Aitken RJ, De Iuliis GN, McLachlan RI. Biological and clinical significance of DNA damage in the male germ line. Int J Androl. 2009;32:46–56.
11. Almeida C, Sousa M, Barros A. Phosphatidilseryne translocation in human spermatozoa from impaired spermatogenesis. Reprod Biomed Online. 2009;19(6):770–7.
12. Alvarez JG, Storey BT. Spontaneous lipid peroxidation in rabbit epididymal spermatozoa. Biol Reprod. 1982;27:1102–8.
13. Alvarez JG, Storey BT. Assessment of cell damage caused by spontaneous lipid peroxidation in rabbit spermatozoa. Biol Reprod. 1984;30:323–32.
14. Alvarez Sedó C, Lavolpe M, Uriondo H, Papier S, Nodar F, Chillik C. Nuclear vacuoles and apoptosis markers in patients with teratozoospermia. JBRA. 2011;15(3):10–4.
15. Alvarez Sedó C, Uriondo H, Serna J, Nodar F, Papier S, Chillik C. Male age and its relationship with sperm apoptosis and lipid peroxidation levels in ART patients. Reproducción. 2012. (In Press).
16. Angelopoulou R, Plastira K, Msaouel P. Spermatozoal sensitive biomarkers to defective protaminosis and fragmented DNA. Reprod Biol Endocrinol. 2007;30(5):36.
17. Antinori M, Licata E, Dani G, Cerusico F, Versaci C, d'Angelo D, et al. Intracytoplasmatic morphologically selected sperm injection: a prospective randomized trial. Reprod Biomed Online. 2008;16(6):835–41.
18. Avendaño C, Franchi A, Taylor S, Morshedi M, Bocca S, Oehninger S. Fragmentation of DNA in morphologically normal human spermatozoa. Fertil Steril. 2009;91(4):1077–84.

19. Avendaño C, Franchi A, Duran H, Oehninger S. DNA fragmentation of normal spermatozoa negatively impacts embryo quality and intracytoplasmic sperm injection outcome. Fertil Steril. 2009;94(2):549–57.

20. Banks S, King SA, Irvine DS, Saunders PT. Impact of a mild scrotal heat stress on DNA integrity in murine spermatozoa. Reproduction. 2005;129:505–14.

21. Barratt CL, Aitken RJ, Björndahl L, Carrell DT, de Boer P, Kvist U, et al. Sperm DNA: organization, protection and vulnerability: from basic science to clinical applications-a position report. Hum Reprod. 2010;25:824–38.

22. Bartke A. Apoptosis of male germ cells, a generalized or a cell type-specific phenomenon? Endocrinology. 1995;136(1):3–4.

23. Bartoov B, Berkovitz A, Eltes F, Kogosowski A, Menezo Y, Barak Y. Real-time fine morphology of motile human sperm cells is associated with IVF-ICSI outcome. J Androl. 2002;23(1):1–8.

24. Berkovitz A, Eltes F, Ellenbogen A, Peer S, Feldberg D, Bartoov B. Does the presence of nuclear vacuoles in human sperm selected for icsi affect pregnancy outcome? Hum Reprod. 2006;21(7):1787–90.

25. Berkovitz A, Eltes F, Lederman H, Peer S, Ellenbogen A, Feldberg B, et al. How to improve IVF-ICSI outcome by sperm selection. Reprod Biomed Online. 2006;12(5):634–8.

26. Boe-Hansen GB, Fedder J, Ersbøll AK, Christensen P. The sperm chromatin structure assay as a diagnostic tool in the human fertility clinic. Hum Reprod. 2006;21(6):1576–82.

27. Bungum M, Humaidan P, Spano M, Jepson K, Bungum L, Giwercman A. The predictive value of sperm chromatin structure assay (SCSA) parameters for the outcome of intrauterine insemination, IVF and ICSI. Hum Reprod. 2004;19:1401–8.

28. Bungum M, Humaidan P, Axmon A, Spano M, Bungum L, Erenpreiss J, et al. Sperm DNA integrity assessment in prediction of assisted reproduction technology outcome. Hum Reprod. 2007;22(1):174–9.

29. Carrell DT, Liu L, Peterson CM, Jones KP, Hatasaka HH, Erickson L, et al. Sperm DNA fragmentation is increased in couples with unexplained recurrent pregnancy loss. Arch Androl. 2003;49:49–55.

30. Cassuto NG, Bouret D, Plouchart JM, Jellad S, Vanderzwalmen P, Balet R, et al. A new real-time morphology classification for human spermatozoa: a link for fertilization and improved embryo quality. Fertil Steril. 2009;92:1616–25.

31. Chohan KR, Griffin JT, Lafromboise M, De Jonge CJ, Carrell DT. Comparison of chromatin assays for DNA fragmentation evaluation in human sperm. J Androl. 2006;27:53–9.

32. Colin A, Barroso G, Gomez-Lopez N, Duran EH, Oehninger S. The effect of age on the expression of apoptosis biomarkers in human spermatozoa. Fertil Steril. 2010;94(7):2609–14.

33. Colonna Worrilow K, Eid S, Matthews J, Pelts E. Multi-site clinical trial evaluating PICSI, a method for selection of hyaluronan-bound sperm for use in ICSI improved clinical results. (Abstract, O17 ESHRE 2010).

34. Cordelli E, Eleuteri P, Leter G, Rescia M, Spanò M. Flow cytometry applications in the evaluation of sperm quality: semen analysis, sperm function and DNA integrity. Contraception. 2005;72(4):273–9.

35. De Iuliis GN, Thomson LK, Mitchell LA, Finnie JM, Koppers AJ, Hedges A, et al. DNA damage in human spermatozoa is highly correlated with the efficiency of chromatin remodeling and the formation of 8-hydroxy-2'-deoxyguanosine, a marker of oxidative stress. Biol Reprod. 2009;81(3):517–24.

36. De Lamirande E, Gagnon C. Capacitacion-associated production of superoxide anion by human spermatozoa. Free Radic Biol Med. 1995;18(3):487–95.

37. Delbes G, Hales BF, Robaire B. Effects of the chemotherapy cocktail used to treat testicular cancer on sperm chromatin integrity. J Androl. 2007;28(2): 241–9.

38. Dirican EK, Özgün OD, Akarsu S, Akm KO, Ercan Ö, Ugürlu M, et al. Clinical outcome of magnetic activated cell sorting of non-apoptotic spermatozoa before density gradient centrifugation for assisted reproduction. J Assist Reprod Genet. 2008;25:375–81.

39. Domínguez-Fandos D, Camejo MI, Ballescà JL, Oliva R. Human sperm DNA fragmentation: correlation of TUNEL results as assessed by flow cytometry and optical microscopy. Cytometry A. 2007;71(12):1011–8.

40. Evenson DP, Darzynkiewicz Z, Melamed MR. Relation of mammalian sperm chromatin heterogeneity to fertility. Science. 1980;210:1131–3.
41. Evenson DP, Jost LK, Corzett M, Balhorn R. Characteristics of human sperm chromatin structure following an episode of influenza and high fever: a case study. J Androl. 2000;21(5):739–46.
42. Evenson DP, Larson KL, Jost LK. Sperm chromatin structure assay: its clinical use for detecting sperm DNA fragmentation in male infertility and comparisons with other techniques. J Androl. 2002;23(1):25–43.
43. Fernández JL, Muriel L, Goyanes V, Segrelles E, Gosálvez J, Enciso M, et al. Simple determination of human sperm DNA fragmentation with an improved sperm chromatin dispersion test. Fertil Steril. 2005;84:833–42.
44. Fraga CG, Motchnik PA, Shigenaga MK. Ascorbic acid protects against endogenous oxidative DNA damage in human sperm. Proc Natl Acad Sci U S A. 1991;88:11003–6.
45. Franco JG Jr, Baruffi RLR, Mauri AL, Petersen CG, Oliveira JBA, Vagnini L. Significance of large nuclear vacuoles in human spermatozoa: implications for ICSI. Reprod Biomed Online. 2008;17(1):42–5.
46. Gharagozloo P, Aitken RJ. The role of sperm oxidative stress in male infertility and the significance of oral antioxidant therapy. Hum Reprod. 2011; 26(7):1628–40.
47. Góngora-Rodríguez A, Fontanilla-Ramírez D. La fragmentación de ADN espermático, influencia sobre las técnicas de reproducción asistida y la calidad embrionaria. Rev Colomb Obstet Ginecol. 2010;61(2).
48. Gorczyca W, Traganos F, Jesionowska H, Darzynkiewicz Z. Presence of DNA strand breaks and increased sensitivity of DNA in situ to denaturation in abnormal human sperm cells: analogy to apoptosis of somatic cells. Exp Cell Res. 1993;207:202–5.
49. Greco E, Romano S, Iacobelli M, Ferrero S, Baroni E, Minasi MG, et al. ICSI in cases of sperm DNA damage: beneficial effect of oral antioxidant treatment. Human Reprod. 2005;20(9):2590–4.
50. Greco E, Scarselli F, Iacobelli M, Rienzi L, Ubaldi F, Ferrero S, et al. Efficient treatment of infertility due to sperm DNA damage by ICSI with testicular spermatozoa. Hum Reprod. 2005;20(1):226–30.
51. Griveau JF, Renard P, Le Lannou D. Superoxide anion production by human spermatozoa as a part of the ionophore-induced acrosome reaction process. Int J Androl. 1995;18:67–74.
52. Haraguchi CM, Mabuchi T, Hirata S, Shoda T, Hoshi K, Akasaki K, et al. Chromatoid bodies: aggresome-like characteristics and degradation sites for organelles of spermiogenic cells. J Histochem Cytochem. 2005;53(4):455–65.
53. Haraguchi CM, Mabuchi T, Hirata S, Shoda T, Tokumoto T, Hoshi K, et al. Possible function of caudal nuclear pocket: degradation of nucleoproteins by ubiquitin-proteasome system in rat spermatids and human sperm. J Histochem Cytochem. 2007;55(6):585–95.
54. Huszar G, Celik-Ozenci C, Cayli S, Zavazcki Z, Hansch E, Vigue L. Hyaluronic acid binding by human sperm indicates cellular maturity, viability and unreacted acrosomal status. Fertil Steril. 2003;79(3):1616–24.
55. Huszar G, Ozkavukcu S, Jakab A, Celik-Ozenci C, Sati GL, Cayli S. Hyaluronic acid binding ability of human sperm reflects cellular maturity and fertilizing potential: selection fo sperm for intracytoplasmic sperm injection. Curr Opin Obstet Gynecol. 2006;18:260–7.
56. Jones R, Mann T, Sherins RJ. Peroxidative breakdown of phospholipids in human spermatozoia: spermicidal effects of fatty acids peroxidatives and protective action of seminal plasma. Fertil Steril. 1976;31:531–7.
57. Kidd SA, Eskenazi B, Wyrobek AJ. Effects of male age on semen quality and fertility: a review of the literature. Fertil Steril. 2001;75:237–48.
58. Knez K, Zorn B, Tomazevic T, Virant-Klu I. The IMSI procedure improves poor embryo development in the same infertile couples with poor semen quality: a comparative prospective randomized study. Reprod Biol Endocrinol. 2011;9:123.
59. Kodama H, Yamaguchi R, Fukuda J, Kasai H, Tanaka T. Increased oxidative deoxyribonucleic acid damage in the spermatozoa of infertile male patients. Fertil Steril. 1997;68(3):519–24.

60. de Lamirande E, Gagnon C. Reactive Oxygen Species (ROS) and reproduction. Adv Exp Med Biol. 1994;366:185–97.
61. Lewis SE, Aitken RJ. DNA damage to spermatozoa has impacts on fertilization and pregnancy. Cell Tissue Res. 2005;322:33–41.
62. Lewis SE, Boyle PM, Mckiney KA. Total antioxidant capacity of the seminal plasma is different in fertile and infertile men. Fertil Steril. 1995;64:868–70.
63. Makhlouf AA, Niederberger C. DNA integrity tests in clinical practice: it is not a simple matter of black and white (or red and green). J Androl. 2006;27(3):316–23.
64. McPherson S, Longo FJ. Chromatin structure-function alterations during mammalian spermatogenesis: DNA nicking and repair in elongating spermatids. Eur J Histochem. 1993; 37(2):109–28.
65. Mehdi M, Khantouche L, Ajina M, Saad A. Detection of DNA fragmentation in human spermatozoa: correlation with semen parameters. Andrologia. 2009;41(6):383–6.
66. Meseguer M, Martínez-Conejero JA, O'Connor JE, Pellicer A, Remohí J, Garrido N. The significance of sperm DNA oxidation in embryo development and reproductive outcome in an oocyte donation program: a new model to study a male infertility prognostic factor. Fertil Steril. 2008;89:1191–9.
67. Meseguer M, Santiso R, Garrido N, Gil-Salom M, Remohí J, Fernandez JL. Sperm DNA framentation levels in testicular sperm samples from azoospermic males as assessed by the sperm chromatin dispersion (SCD) test. Fertil Steril. 2009;92(5):1638–45.
68. Morris ID, Ilott S, Dixon L, Brison DR. The spectrum of DNA damage in human sperm assessed by single cell gel electrophoresis (comet assay) and its relationship to fertilization and embryo development. Hum Reprod. 2002;17:990–8.
69. Moskovtsev SI, Willis J, White J, Mullen JB. Sperm DNA damage: correlation to severity of semen abnormalities. Urology. 2009;74(4):789–93.
70. Moskovtsev SI, Jarvi K, Mullen JBM, Cadesky KI, Hannam T, Lo KC. Testicular spermatozoa have statistically significantly lower DNA damage compared with ejaculated spermatozoa in patients with unsuccessful oral antioxidant treatment. Fertil Steril. 2010;93(4):1142–6.
71. Moustafa MH, Sharma RK, Thornton J, Mascha E, Abdel-Hafez MA, Thomas AJ Jr, et al. Relationship between ROS production, apoptosis and DNA denaturation in spermatozoa from patients examined for infertility. Hum Reprod. 2006;19(1):129–38.
72. Oehninger S, Morshedi M, Weng SL, Taylor S, Duran H, Beebe S. Presence and significance of somatic cell apoptosis markers in human ejaculated spermatozoa. Reprod Biomed Online. 2003;7(4):469–76.
73. O'Flaherty C, Vaisheva F, Hales BF, Chan P, Robaire B. Characterization of sperm chromatin quality in testicular cancer and Hodgkin's lymphoma patients prior to chemotherapy. Hum Reprod. 2008;23(5):1044–52.
74. Oliveira JBA, Massaro FC, Mauri AL, Petersen CG, Nicoletti APM, Baruffi RLR, et al. Motile sperm organelle morphology examination is stricter than Tygerberg criteria. Reprod Biomed Online. 2009;18(3):320–6.
75. Oliveira JBA, Massaro FC, Baruffi RLR, Mauri AL, Petersen CG, Silva LFI, et al. Correlation between semen analysis by motile sperm organelle morphology examination and sperm DNA damage. Fertil Steril. 2010;94(5):1937–40.
76. Oliveira JBA, Petersen CG, Massaro FC, Baruffi RLR, Mauri AL, Silva LFI, et al. Motile sperm organelle morphology examination (MSOME): intervariation study of normal sperm and sperm with large nuclear vacuoles. Reprod Biol Endocrinol. 2010;8:56–62.
77. Ollero M, Gil-Guzman E, Lopez MC, Sharma RK, Agarwal A, Larson K, et al. Characterization of subsets of human spermatozoa at different stages of maturation: implications in the diagnosis and treatment of male infertility. Hum Reprod. 2001; 16(9):1912–21.
78. Palan P, Naz R. Changes in various antioxidant levels in human seminal plasma related to immunoinfertility. Arch Androl. 1996;36(2):139–43.
79. Parmegiani L, Cognigni GE, Bernardi S, Troilo E, Ciampaglia W, Filicori M. "Physiologic ICSI": hyaluronic acid (HA) favors selection of spermatozoa without DNA fragmentation and with normal nucleus, resulting in improvement of embryo quality. Fertil Steril. 2010;93(2):598–604.

80. Parmegiani L, Cognigni GE, Ciampaglia W, Pocognoli P, Marchi F, Filicori M. Efficiency of hyaluronic acid (HA) sperm selection. J Asist Reprod Genet. 2010;27:13–6.

81. Print CG, Loveland KL. Germ cell suicide: new insights into apoptosis during spermatogenesis. Bioessays. 2000;22(5):423–30.

82. Rawe V, Alvarez SC, Uriondo H, Papier S, Miasnik S, Nodar F. Separación magnética por columnas de Anexina V: filtrado molecular para la selección de espermatozoides no apoptóticos. Reproducción. 2009;24:104–14.

83. Rawe VY, Boudri H, Alvarez SC, Carro M, Papier S, Nodar F. Healthy baby born after reduction of sperm DNA fragmentation using cell sorting before ICSI. Reprod Biomed Online. 2010;20:320–3.

84. Rawe V, Sueldo CM, Blanco L, Sueldo CE. Is sperm DNA fragmentation a hidden male factor in human infertility?. Fertil Steril. 2010;94(4):34–5.

85. Rougier N, Uriondo H, Nodar F, Papier S, Sueldo C. Alvarez Sedó. Changes in DNA fragmentation in preparation for ICSI at variable time periods. (Abstract) ASRM Annual Meeting San Diego; 2012.

86. Said TM, Grunewald S, Paasch U, Glander H, Baumann T, Kriegel C, et al. Advantage of combining magnetic cell separation with sperm preparation techniques. Reprod Biomed Online. 2005a;10(6):740–6.

87. Said TM, Grunewald S, Paasch U, Glander HJ, Baumann T, Kriegel C, et al. Advantage of combining magnetic cell separation with sperm preparation techniques. Reprod Biomed Online. 2005b;10(6):740–6.

88. Sakkas D, Alvarez JG. Sperm DNA fragmentation: mechanisms of origin, impact on reproductive outcome, and analysis. Fertil Steril. 2010; 93(4):1027–36.

89. Sakkas D, Mariethoz E, Manicardi G, Bizzaro D, Bianchi PG, Bianchi U. Origin of DNA damage in ejaculated human spermatozoa. Rev Reprod. 1999; 4(1):31–7.

90. Sakkas D, Moffatt O, Manicardi GC, Mariethoz E, Tarozzi N, Bizzaro D. Nature of DNA damage in ejaculated human spermatozoa and the possible involvement of apoptosis. Biol Reprod. 2002;66(4):1061–7.

91. Sakkas D, Seli E, Bizzaro D, Tarozzi N, Manicardi GC. Abnormal spermatozoa in the ejaculate: abortive apoptosis and faulty nuclear remodeling during spermatogenesis. Reprod Biomed Online. 2003;7(4):428–32.

92. Seli E, Gardner DK, Schoolcraft WB, Moffatt O, Sakkas D. Extent of nuclear DNA damage in ejaculated spermatozoa impacts on blastocyst development after in vitro fertilization. Fertil Steril. 2004;82:378–83.

93. Sergerie M, Laforest G, Bujan L, Bissonnette F, Bleau G. Sperm DNA fragmentation: threshold value in male fertility. Hum Reprod. 2005;20(12):3446–51.

94. Shamsi MB, Venkatesh S, Tanwar M, Singh G, Mukherjee S, Malhotra N, Kumar R, Gupta NP, Mittal S, Dada R. Comet assay: a prognostic tool for DNA integrity assessment in infertile men opting for assisted reproduction. Indian J Med Res. 2010;131:675–81.

95. Sharma RK, Agarwall A. Role of reactive oxygen species in male infertility. J Urol. 1996;48:835–50.

96. Sharma RK, Pasqualotto FF, Nelson DR. The reactive oxygen species-total antioxidant capacity score is a new measure of oxidative stress to predict male infertility. Hum Reprod. 1999;14:2801–7.

97. Sharma RK, Sabanegh E, Mahfouz R, Gupta S, Thiyagarajan A, Agarwal A. TUNEL as a test for sperm DNA damage in the evaluation of male infertility. Urology. 2010;76(6):1380–6.

98. Singh NP, Muller CH, Berger RE. Effects of age on DNA double-strand breaks and apoptosis in human sperm. Fertil Steril. 2003;80(6):1420–30.

99. Smit M, van Casteren NJ, Wildhagen MF, Romijn JC, Dohle GR. Sperm DNA integrity in cancer patients before and after cytotoxic treatment. Hum Reprod. 2010;25(8):1877–83.

100. Smith R, Kaune H, Parodi D, Madariaga M, Morales I, Ríos R, et al. Extent of sperm DNA damage in spermatozoa from men examined for infertility. Relationship with oxidative stress. Rev Med Chil. 2007;135(3):279–86.

101. Tamburrino L, Marchiani S, Montoya M, Elia Marino F, Natali I, Cambi M, Forti G, Baldi E, Muratori M. Mechanisms and clinical correlates of sperm DNA damage. Asian J Androl. 2012;14(1):24–31.
102. Tarozzi N, Nadalini M, Bizzaro D, Borini A, et al. Sperm-hyaluronan binding assay: clinical value in standard IVF under the Italian Law. Reprod Biomed Online. 2009;19(3):35–43.
103. Uriondo H, Alvarez Sedó C, Virginia GM, Frazer P, Serna J, Nodar F. Correlation between phosphatidylserine externalization and sperm apoptosis in infertile men. Reproducción. 2011;26(3):111–6.
104. Vagnini L, Baruffi RL, Mauri AL, Petersen CG, Massaro FC, Pontes A, et al. The effects of male age on sperm DNA damage in an infertile population. Reprod Biomed Online. 2007;15(5):514–9.
105. Virro MR, Larson-Cook KL, Evenson DP. Sperm chromatin structure assay parameters are related to fertilization, blastocyst development, and ongoing pregnancy in in vitro fertilization and intracytoplasmic sperm injection cycles. Fertil Steril. 2004;81:1289–95.
106. Wu GJ, Chang FW, Lee SS, Cheng YY, Chen CH, Chen IC. Apoptosis-related phenotype of ejaculated spermatozoa in patients with varicocele. Fertil Steril. 2009;91(3):831–7.
107. Ye H, Huang G, Gao Y, Liu DY. Relationship between human sperm-hyaluronan binding assay and fertilization rate in conventional in vitro fertilization. Hum Reprod. 2006;21(6):1545–50.
108. Zini A, Sigman M. Are tests of sperm DNA damage clinically useful? Pros and cons. J Androl. 2009;30:219–29.
109. Zini A, Azhar R, Baazeem A, Gabriel MS. Effect of microsurgical varicocelectomy on human sperm chromatin and DNA integrity: a prospective trial. Int J Androl. 2010;34:14–9.